Tuning Innovation with Biotechnology

Tuning Innovation with Biotechnology

Dong Hwa Kim

Pan Stanford Publishing

Published by

Pan Stanford Publishing Pte. Ltd.
Penthouse Level, Suntec Tower 3
8 Temasek Boulevard
Singapore 038988

Email: editorial@panstanford.com
Web: www.panstanford.com

British Library Cataloguing-in-Publication Data
A catalogue record for this book is available from the British Library.

Tuning Innovation with Biotechnology

ISBN 978-981-4745-35-2 (Hardcover)
ISBN 978-1-315-36458-2 (eBook)

Printed in the USA

Contents

Preface

If we compare today's life pattern with that of the 1980s, we can see that a lot of things have changed and our life style has evolved not only in terms of knowledge but also with respect to dependence on the external factors. That is, the domain of our knowledge can be expanded and depends on external entities.

Now if we broaden our domain from the computer system to the machines all around us, this domain of required knowledge is changing and most of the jobs that were associated with human beings will depend on systems such as robots and ICT. Therefore, it is clear that innovation for these domains of required knowledge is needed—not for making computers efficient but for empowering them to do things.

Around the world, from a technology perspective, artificial intelligence plays an important role as a key technology in pattern recognition, automobile, contents, decision-making, cellular phones, networking, robots, etc.

From a societal perspective, the engineers for such systems should possess more flexible ideas in terms of knowledge representation, pedagogical strategies, and adaptation technologies. All of these factors can contribute to teaching, learning, creative ideas, development, and science education.

In the human history, there have been several revolutions in technology. The first industrial revolution was the revolution in agriculture with cultivation technology and the second revolution was the machine power revolution marked by steam power, electricity, and the internal combustion engine. Some researchers would term the electric power revolution the third revolution. However, I would like to assert it as the continuation of the steam power revolution. The third revolution, i.e., the information revolution, began in the 1960s and was based on digital technology, personal computers, and the development of the Internet.

Recently, there has been a mention of the fourth revolution. The fourth revolution is much more complicated with AI, Internet, mobile, design, contents, and their fusion. It will have a big influence on areas such as driverless cars, smart robotics, mobile technology, display methods, 3D printing, and materials that are lighter and tougher.

In the past, it always took time to feel the full effects of a technological change. Today, many advances are at their starting point. However, the fourth revolution has a very high speed and is further complicated through convergence.

One of the most profound impacts of the fourth industrial revolution will be on what jobs people have and what skills they will need to succeed. Of course, all industrial facilities will transition and jobs will also be changed substantially.

Of course, the fruits of an economy dominated by artificial intelligence and related smart technology will be redistributed and we will have more leisure time (I would like to say, "smart life") and machines will do all the work.

Therefore, those who are not prepared will lose. We can also expect many jobs to disappear, and new jobs will be created by these technologies. I am sure the future core technology will be artificial intelligence or related technology.

This book deals with evolving intelligence systems and their use in immune algorithm (IM), particle swarm optimization (PSO), bacterial foraging (BF), and hybrid intelligent system to improve plants, robots, etc. It discusses the motivation behind research on and background of evolving intelligence systems and illustrates IM-based approach for parameter estimation required for designing an intelligent system. It approaches optimal intelligent tuning using a hybrid genetic algorithm–particle swarm optimization (GA-PSO), illustrates hybrid GA-PSO for the intelligent tuning of a vector system, and describes hybrid systems consisting of GA and BF for an intelligent system. Finally, it describes potential applications, such as an industrial intelligent system, an intelligent network, emotional robotics, and emotional ICT. The book will be a useful reference for researchers and graduate students in computer science or related disciplines such as computer engineering, philosophy, cognitive science, and psychology.

Finally, the author would like to thank Pan Stanford Publishing Pte. Ltd. for making this book a reality.

Chapter 1

Background

This chapter explains the research motivation for bio-based tuning issues such as particle swarm optimization (PSO), immune algorithm, bacterial foraging, and its hybrid system for tuning.

1.1 Overview

In recent years, combined learning-based artificial intelligence (AI) such as neural networks (NNs), fuzzy controls, genetic algorithms, bacteria foraging (BF), particle swarm optimization, and immune network structures have been of interest because of their robustness and flexibility against a dynamically changing system or complex system. At the same time, conventional artificial intelligent systems based on a functional decomposition and leading to the so-called "sense-model-plan-action" cycle have been criticized on many grounds over the past decade [8, 30]. The new, soft-computing approaches are used extensively in such diverse applications as fault prediction, fault diagnosis, supervisory control, energy management, production management, and software engineering, among others [16].

Over the past decade, significant advances have been made in two areas of computational intelligence: fuzzy logic (FL) and

Tuning Innovation with Biotechnology
Dong Hwa Kim

ISBN 978-981-4745-35-2 (Hardcover), 978-1-315-36458-2 (eBook)
www.panstanford.com

neural networks [1, 5, 6, 10, 13]. For instance, there has been a considerable interest in the past few years in exploring the applications of fuzzy neural network (FNN) systems, which combine the capability of fuzzy reasoning to handle uncertain information and the capability of artificial networks to learn from processes [10–13], to deal with nonlinearities and uncertainties of control systems [14, 15].

It is of interest to research in computer science and engineering that biological information processing systems such as human beings have many functions useful to model intelligent behavior and are expected to provide various ideas feasible to engineering fields, especially intelligent control or robotics [6, 8, 45]. Biological information in living organisms can be mainly classified into the following systems: brain nervous system, genetic systems, endocrine systems, particle swarm optimization, bacteria foraging, and immune systems. Among these, certain systems have already been applied to engineering fields by modeling, control, fault tolerance, identification, and parameter estimation: These include especially the neural network and evolutionary algorithms [12, 22, 23]. That is, the above two approaches have been widely used in various practical applications. Especially, the technology of fuzzy and neural control in process field has grown very rapidly in recent years. Neural network control schemes of many kinds have been proposed and used in a range of technical applications. This growth has been fueled by the renewed attention to neural networks and their application in intelligent control areas. Fuzzy control is the other control area that has attracted similar attention in the past decade. The rationale for using neural control or fuzzy logic control systems is related to difficulties, which are commonly experienced by control engineers in real-world applications.

It is generally difficult to accurately represent a complex process or system by a mathematical model or by a simple computer model. Even if the model itself is tractable, control of the process using a "hard" (nonsoft or crisp) control algorithm might not provide satisfactory performance. Furthermore, it is commonly known that the performance of some industrial processes can be considerably improved through the high-level control actions made by an experienced and skilled operator, and these actions normally cannot be formulated by crisp control algorithms [19–21].

From the control theory point of view, the control engineers are indispensable in selecting and designing control system for new plants. Their work is equally useful in analyzing and adjusting existing plants so that their optimum performance is achieved. Generally, the need for improved control performance is caused by the following requirements:

- improvement in product quality
- increase in process yield or production rate
- increased safety for personnel and equipment
- increased economic efficiency of operation resulting in reduced in energy, materials, time

If the process model itself is inaccurate or insufficient, model-based control cannot provide satisfactory results. Even when an accurate model is provided, if the parameter values are not fully known, then appropriate estimates have to be made before operation. Therefore, crisp control algorithms based on incomplete information usually will not give satisfactory results. Generally speaking, a primary purpose of classical feedback control is to increase the robustness of the control system, i.e., to increase the performance of the system when there is uncertainty such as model errors, unknown disturbances, and noise. Crisp-based adaptive and robust control techniques are designed to cope with uncertainties due to large variations in parameter values, environmental conditions for control information, and signal inputs. However, the region of operability of the control system will be restricted, although it will be considerably large compared to nonrobust classical control systems [7, 35]. In complex processes in practice, the range of uncertainty may be substantially larger than can be tolerated by crisp algorithms of adaptive and robust control. What are known as "intelligent" control techniques are useful here; one example is evolutionary control. Herein, it is important for us to understand neural networks, fuzzy systems, and evolutionary approaches to control such as genetic algorithms, and immune networks [24, 29, 35, 36, 74–83].

The control technology in industrial fields, resulting in increased growth in size and complexity of process plants, has given rise to the need for improved control of such plants. Increasing economic pressure within the industries is forcing managements to look for more effective means to design and operate even more efficient process plants. The greatest recent advance in the process industries has been the widespread application of the techniques of automatic control and intelligent control in the design of new plants and in the improvement of the existing ones.

This application is popularly known as modern control or intelligent control, which means to control automatically or intelligently the operation of an apparatus, process, or system through the duplication by computer means of the activities of living creatures—in short, control by computer-based intelligent algorithms instead of men [49, 106, 107, 117]. The effects of intelligent automation are currently being amplified by the use of computers as both tools and controllers in the design of control systems and as integral components of such systems. Therefore, since it is important to obtain the required response from the given plant, intelligent approaches are increasingly accepted for control and information processes in industry [17, 23, 24, 35, 38, 64]. In particular, successful use of evolutionary methods such as GAs depends heavily on coding strategies for underlying applications [97, 99, 100, 103, 106, 107]. For instance, the coding scheme for a fuzzy inference system (FIS) refers to the way of arranging the parameters of the FIS into a bit-string representation (or chromosome) such that the representation preserves certain useful properties after recombination specified by the genetic operators like crossover and mutation. The standard crossover and/or mutation operators may produce incomplete or ill-defined FISs. Herein, It is necessary for the researchers to develop many kinds of coding that can lead to more efficient algorithms and deal with these problems through simulation or experiments on target system.

1.2 PID Controller Tuning in Automatic Control System

Even though control theory has been developed significantly, the proportional-integral-derivative (PID) controllers are used for a wide range of process control, motor drives, magnetic and optic memories, automotive, flight control, instrumentation, etc. In industrial applications, more than 90% of all control loops are PID type [6, 19, 20, 25, 40]. Integral, proportional, and derivative feedback is based on the past (I), present (P), and future (D) control conception. Over the last 50 years, several methods for determining PID controller parameters have been developed for stable processes that are suitable for auto-tuning and adaptive control [20, 21, 31–34, 36, 38–41, 64, 74, 78].

The tuning of these controllers for plants with over-damped step responses is well studied, and many tuning formulas are given in the literature: the Ziegler–Nichols formula [2], the advanced Ziegler–Nichols formula [31, 35], the Cohen–Coon formula, the internal model control design formula, the integral absolute error optimum formula, the integral squared error optimum formula, the integral time-weighted square error optimum formula [9], the integral time-weighted absolute error optimum formula [7], and the gain and phase margins (GPM) formula [27], and intelligent approaches [34, 75]. The effectiveness of the PID controller is, however, not well understood for plants with complex tuning parameter. The classical textbook example of a system with this property is the mechanical system [19].

Some heuristic knowledge on the tuning of the PID controller for plant with complex tuning parameters can be found in [40]. In practice, one may have to tune the PID controller manually through trial-and-error procedures for a plant with under-damped step response. The situation is clearly very unlike the case of the plant with over-damped step response where many well-known formulas exist [41, 98]. Gain and phase margins have always served as important measures of robustness. It is also known from classical control that phase margin is related to the damping of the system and therefore can also serve as a performance measure [35]. Controller designs to satisfy the gain margin and phase margin criteria are not new [35, 37]. However, they are normally solved using numerical methods on computers or graphically by trial and error using Bode plots. Both these approaches are not very suitable for adaptive tuning.

Some researchers show how certain simple approximations can be introduced into the calculations so that neither numerical methods nor graphical methods have to be used for intelligent control [27, 36]. Inverse function mapping for tuning was performed by fuzzy neural networks (FNN). Namely, the FNNs identify the relationship between GPM and PID controllers difficult to describe analytically. Such relationships can be formulated analytically for a general linear system [93, 95]. For solving the problem, the first obstacle is the difficulty in finding the stabilizing region of PID controllers.

A simplified approach based on approximation has been adopted to treat this problem. To extend the applications to general plant, it is important to introduce the FNN for identifying the relationship between PID controllers and GPM. It is used to automatically tune the PID controller parameters for GPM specifications so that neither numerical methods nor graphical methods need to be used. Along with these approaches, a number of approaches have been proposed to implement mixed control structures that combine a PID controller with fuzzy logic. The first such attempt combined a PID controller and a fuzzy controller in parallel. To do so, the fine-tuning of the fuzzy controller could be replaced with the PID control [12, 17]. In the hierarchical approach of its authors, a fuzzy system supervised the PID controller in order to improve the dynamic response of the closed-loop system.

Currently, process and control engineers can successfully regulate a majority of industrial processes by meeting various specifications under consideration. Analytical studies on transient response, stability, and reliability guarantee dynamic performance of the PID controllers in some nominal operating conditions. However, the capability of the PID controllers is significantly reduced when they are applied to systems with nonlinearities such as saturation, relay, hysterisis, and dead zone. In this case, intelligent tuning systems can improve the performance of the PID control systems.

In spite of their individual philosophies and structural difference, all of these approaches share the same difficulty of tuning rule generation. Tuning rules in a mixed control structure must be different according to the plant and the conditions in which it is operated. Since the generation of tuning rules is a difficult and time-consuming procedure, it is required to have a systematic method for constructing appropriate rules.

Some papers demonstrate that GA is an efficient and robust tool for generating fuzzy rules in fuzzy-PID hybrid control structures [24, 35]. That is, they suggest that the GA can construct a set of fuzzy rules that optimize multiple criteria. An important observation that the rules searched by GA are randomly scattered is made and a solution to this problem is provided by including a smoothness cost in the objective function. The results suggest that various control schemes such as fuzzy-PID hybrid and NN-PID hybrid control schemes can be considered in the context of tuning the PID controllers. A general view on the set-point modification and the gain modification is given and several hybrid structures are shown to be special cases of tuning. Here, intelligence-based tuning approach for the control system is needed in cases of a plant.

1.3 Evolutionary-Based Intelligent Control

1.3.1 Immune Algorithm–Based Intelligent Control

The immune network theory has been first proposed by Jerne [47] as a hypothesis such as a novel viewpoint of lymphocyte activities, natural antibody production, pre-immune repertoire selection, tolerance and self/non-self-discrimination, memory and the evolution of immune systems. The immune system is composed of a regulated network of cells and molecules that recognize one another even in the absence of antigens, and it is generally defined as an enormous and complex network of paratopes that recognize sets of idiotopes, and of idiotopes that are recognized by sets of paratopes; thus it could recognize as well as be recognized. The cells in immune network can respond either positively or negatively to the recognition signal (antigen or other immune cell or molecule). A positive response would result in cell proliferation, cell activation, and antibody secretion, while a negative response would lead to tolerance and suppression [42–59].

Through this activation, the immune system can protect our bodies from attack of foreign matters (antigens) that enter the blood stream. Theoretical biologists who are interested in modeling the immune system have found it useful to apply to engineering areas as documented in a diverse literature. Other researchers interested in immunology are engineers who try to mimic the immune system in order to create better artificial systems. For example, the strong parallels between immunology and computer security allow us to abstract principles and mechanisms that enable design of better computer security systems and higher order control engineering problems [53, 64, 77, 78, 99].

As the result, in recent years, there has been a growing interest in using evolutionary algorithms for many applications in control systems [10, 18, 24], because a control system is often subject to changes due to various sources of uncertainty. For instance, a small deviation from an operating condition may have disastrous effects on a system. The performance characteristics such as stability and robustness may need to be re-examined as the degrees of complexity and uncertainty increase. Resolving uncertainty in systems design in general and in control systems in particular has been an important issue.

The purpose of introducing immunological concepts into engineering and other computational intelligence paradigms (GA, fuzzy, and neural network) is to use the locally characteristic information for seeking the ways and means of finding the optimal solution when dealing with difficult problems or complex variable systems. That is, immune approaches utilize local information to intervene in the globally parallel processes and restrain or avoid repetitive and useless work during the course, so as to overcome the blind actions of crossover and mutation. There are two aspects of research on immune systems: the general characteristic of an immune network and the efficient application of immunological concepts to the given problem being solved.

There are several immune network models presented in the literature (models are described in the appendix). Most of them are based on a set of differential equations to describe the dynamics of the network cells and molecules [46, 47]. There are two main interactions: stimulation function and suppression function between different types of elements. They lead naturally to the network connectivity pattern and dynamics [44–56].

Based on this network system, the immunized computational algorithm (ICA) that uses the immune system metaphor along with computational (both hard and soft computing) techniques to attempt to reproduce the robustness and adaptability of a biological immune system is suggested by Krishnakumar [42] and Krishnakumar and Neidhoefer [50]. There is also an increase in hybrid structures whose building blocks, mimicking the immune building blocks, are composed of artificial neural networks, fuzzy systems, and evolutionary algorithms. Their immune system is focused on the clonal selection principle, together with the affinity maturation of the immune response [74, 75]. Some results depict their viewpoint of the trade-off between the immune system and the intelligent control theory [48, 50, 74]. This section is intended to be a comprehensive overview to the immune system for the next chapter, which is going to explain much application in immunology.

1.3.2 PSO–Based Intelligent Control and Hybrid System

As particle swarm optimization (PSO) is an algorithm for finding optimal regions of complex search spaces through interaction of individuals in a population of particles, it usually uses a system of two iterative equations, one for the positions and the other for the velocities of the particles, with several parameters. It gives more "freedom" to the system, but it is also then quite difficult to find the best parameters values to each case.

This kind of algorithm is still largely experimentally studied: It is indeed extremely efficient [106, 108]. In 1998, a mathematical explanation, introducing a five-dimensional space of states and some constriction coefficients for sure convergence, was found [8], but as there was still no sure way to choose a priori the best parameters, their parameters (the velocity/position coefficient) were randomly tuned or selected at each time step while applying to the engineering area.

Some results of the particle swarm optimization suggest methods for altering the original algorithm in ways that eliminate problems and increase the optimization power of the particle swarm [105]. The classical particle swarm optimization is a powerful method to find the minimum of a numerical function, on a continuous definition domain [107]. In order to further research both the power to engineering application and the limits of this approach, it is necessary to examine in detail how it can be used to solve the complex system or the well-known PID tuning and to estimate the particles current, the previous positions, and function values in the PSO algorithm [108].

There are different versions of particle swarm optimization algorithms, but we need to study the following from an engineering or informational point of view: what kind of information (control) each particle has access to, and how it uses it. To illustrate this, we study two approaches, a PSO-based control problem and the hybrid system based on PSO [56–60]. Of course, schemes in this context are not enough for using information and control area and are not at all costless for real engineering. Since its efficiency depends not only on the results but also on the approach to use it, the efficiency versus information obtained may not always be satisfying. Therefore, we need to study further from this point.

1.3.3 Bacteria Foraging–Based Intelligent Control and Hybrid System

As natural selection tends to eliminate animals with poor foraging strategies through methods for locating, handling, and ingesting food and favor the propagation of genes of those animals that have successful foraging strategies, they are more likely to apply reproductive success to have an optimal solution [110, 112]. The basic principle of natural selection is to search for and obtain nutrients in a way that maximizes the ratio between the energy obtained and the time spent foraging or maximizes the long-term average rate of energy intake. Evolution optimizes the foraging strategies, since animals that have poor foraging performance do not survive. Generally, a foraging strategy involves finding a patch of food. There are predators and risks, energy required for travel, and physiological constraints (sensing, memory, cognitive capabilities).

Foraging strategies can be modeled as optimal policies and the results can be applied in engineering using, for instance, dynamic programming, optimal control, and optimal tuning [115–117]. Some animals forage as individuals and others forage as groups. To perform social foraging, an animal needs communication capabilities, it can gain advantages in that it can exploit essentially the sensing capabilities of the group, the group can gang up on large prey, individuals can obtain protection from predators while in a group, and in a certain sense the group can forage with a type of collective intelligence. Scientifically, scientists in the field of foraging theory have led such evolutionary principles to hypothesize that it is appropriate to model the activity of foraging as an optimization process: A foraging animal takes actions to maximize the energy obtained per unit time spent foraging, in the face of constraints presented by its own physiology such as sensing and cognitive capabilities and environment (e.g., density of prey, risks from predators, physical characteristics of the search area) [116]. Evolution has balanced these constraints and essentially engineered what is sometimes referred to as an optimal foraging policy. Optimization models are also valid for social foraging where groups of animals communicate to forage cooperatively. Such social foragers include birds, bees, fish, ants, wild beasts, and primates [114, 115]. As mentioned above, foraging can be modeled as an optimization process where an animal seeks to maximize the energy obtained per unit time spent foraging. From an engineering perspective, ends of such a spectrum are interesting in application or study of principle.

1.3.4 Hybrid System–Based Artificial Intelligence

Hybrid systems are powerful systems from the point of view of knowledge engineering and cognitive science. Computing processing using the advantage of each module in a hybrid system may provide, for instance, learning, robustness, graded plausible interpretation to system, easy understanding, simple representation, and efficiency coding in smaller domains. In particular, novel classes of hybrid neural, fuzzy techniques, genetic algorithms, bacteria foraging, and particle swarm optimization or so support for analyzing tool, new dynamic principles and methodologies should be developed in order to advance the theory and algorithms for hybrid system development. This approach could be examined and further developed, for instance, in the scenario of building a natural system or new novel based on recurrent networks and symbolic hybrid transducers [111–113].

Hybrid systems can also dynamically use or apply to engineering. Furthermore, the number of hybrid systems can vary dynamically over time. If the best possible hybrid agent may be available, the next best choice has to be found in order to supply a best possible computing or intelligence like human being [100, 101, 109, 112, 118, 119]. This dynamic unrestricted message routing task is new, and hybrid techniques have not yet been examined for noisy dynamic systems. Hybrid techniques are chosen for this topic since hybrid networks support robustness and learning in noisy unrestricted real-world texts or engineering area and since easy dynamic models support architecture control and exception handling. Herein, we show the principle of a hybrid system application to control system [103–105, 110]. A detailed description is presented in Chapter 4.

1.4 Relationship in Research Topic

This book is organized as follows. In Chapter 1, Section 1.1 presents an overview of intelligence and Section 1.2 discusses the role of a PID controller in an automatic system and necessity and motivation of intelligent tuning. Section 1.3 describes needs of evolutionary-based intelligent control such as immune algorithm, particle swarm optimization, and bacteria foraging.

In Chapter 2, optimal nonlinear parameter estimation of induction motor using immune algorithm is suggested for induction motor control. Chapter 3 suggests a hybrid system (GA-PSO) and shows the hybrid system using particle swarm optimization and genetic algorithm, and intelligent PID controller tuning. Chapter 4 describes GA-PSO-based PI controller tuning for vector control of induction motor. Chapter 5 introduces the novel hybrid system based on GA and bacteria foraging. In addition, for obtaining a better time-response performance, a PID controller tuning to meet the control specification is described.

Final conclusions and directions for further work are given in Chapter 7.

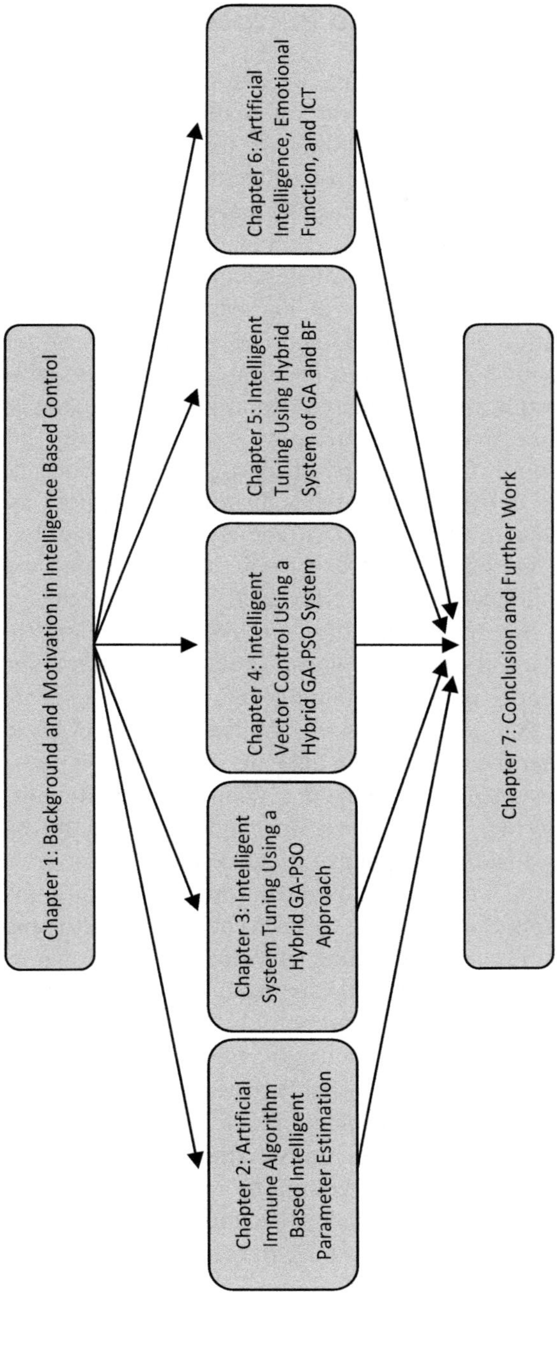

Figure 1.1 Topic organization of this book.

1.5 Further Work and Perspective

Intelligent system (IS) is one of the most widespread in the contemporary computer world no matter what the domain of application is, or which discipline we use for the analysis of this domain. Intelligent systems are expected to work well in many different environments and usually have perception, knowledge, decision-making capabilities, and behaviors to follow the decisions. These properties of intelligent systems allow them to maximize the probability of success even if full knowledge of the situation is not available. These functions of intelligent systems cannot be considered separately from the environments in control system and the concrete situation including the goal.

To ensure that the control system reaches the goals, we need to choose functions (control algorithms) such as fuzzy logic, neural networks, genetic algorithms, biological approach; hybrid intelligent system, with some methodology to control system. The intelligent control system still has a long way to go in the real industrial field. However, it can be believed that the answer lies in the building of intelligence algorithm from the ground up through a hybrid intelligent system such as a neural network, biological intelligence, and other evolutionary algorithms. The possibilities of intelligent control system are endless. However, we should also be concerned if we need to evaluate the system's intelligence from an outside perspective (e.g., considering it as a black box and analyzing its behavior) or from an inside perspective (e.g., analyzing its architecture and identifying structures, organizations, computing processes, and algorithms that are able to be treated as performance of its intelligence). From these points of view, the level of intelligence of a control system is directly associated to the algorithms that the controller is able to compute in its control architecture. Therefore, the determination of the types of intelligence algorithms involved in the architecture, besides the architecture itself, will lead to an evaluation of the overall intelligence of the system. Intelligence control should have the ability of something (system, device, being) to evaluate the achievement of a control purpose, and evaluate algorithm to drive its further behavior in order to let the purpose to realize.

Chapter 2

Artificial Immune Algorithm–Based Intelligent Parameter Estimation

This chapter presents immune algorithm–based parameter estimation for intelligent and its hybrid system.

2.1 Overview

Even though many kinds of tuning schemes have been developed significantly, it is a challenge for controller tuning technology to explore novel control strategies and philosophies for complex industrial processes [14, 15, 19]. The application of intelligent system technologies in industrial control has been developed into an emerging technology, so-called "artificial intelligent system" [17, 18]. This technology is highly multidisciplinary and rooted in system control, analysis, artificial intelligence, information and signal processing, computer software, and production [13, 25, 28].

Chronologically, fuzzy logic was the first technique of intelligent system. Neural, neuro-fuzzy and evolutionary systems and their derivatives followed later [2, 5, 10, 11]. Each technique is offering new possibilities and making intelligent system in industrial application [10, 23, 24].

On the other hand, biological information-processing systems such as human beings have many interesting functions and are expected to provide various feasible ideas to engineering areas,

Tuning Innovation with Biotechnology
Dong Hwa Kim

ISBN 978-981-4745-35-2 (Hardcover), 978-1-315-36458-2 (eBook)
www.panstanford.com

especially intelligent system or robotics [6, 12, 45]. However, little attention has been paid to the applications of some other systems, such as, for instance, immune algorithm in engineering, despite their very important characteristics and model for intelligent system.

The artificial immune system (AIS) implements a learning technique inspired by the human immune system, which is a remarkable natural defense mechanism that learns about foreign substances, However, the immune system has not attracted the same kind of interest from the computing field as the neural operation of the brain or the evolutionary forces used in learning classifier systems [5, 12, 26, 28, 42, 60].

The immune system is a rich source of theories and as such can act as an inspiration for computer-based solutions. Other areas of the interest relating to the characteristics of the immune system are listed below [46–59]:

- The learning method of the immune system is a distributed system with no central controller, since the immune system is distributed and consists of an enormous number and diversity of cells throughout our bodies.
- The immune system has a naturally occurring event-response system which can quickly adapt to changing situations and shares the property with the central nervous system that a definite recognition can be made with a stimulus.
- The immune system possesses a self-organizing and distributed memory. Therefore, it is thus adaptive to its external environment and allows a parallel distributed processing (PDP) network to complete patterns against the environmental situation.
- The correct functioning of the immune system is insensitive to the fine details of the network connections, since a significant part of the immune system repertoire is generate by somatic mutation processes.

In particular, immune system has various interesting features such as immunological memory, immunological tolerance, pattern recognition, immunology computing, and others, when viewed from the engineering viewpoint [38, 51, 74, 77].

Brooks, a pioneer of the several robotic approaches, has presented subsumption architecture for behavior arbitration of autonomous robots [45]. He has argued that intelligence should emerge from mutual interactions among competence modules (i.e., simple behavior/action), and interactions between a robot and its environment. However, the behavior-based artificial intelligence (AI) still has the following questions: "how do we construct an appropriate method for system? or how do we organize appropriate for the given engineering aspects?"

Recently the artificial immune system (AIS) research area has been interesting, since it has particularly various aforementioned interesting features. Therefore, it can play important role to maintain its own system against dynamically changing environments and expected to provide, through the interaction among lymphocytes and/or antibodies, a new methodology suitable for dynamic problems dealing with unknown hostile environments rather than static problem.

From the above facts, some researchers [87, 117, 118] who particularly focused on the similarities between the behavior arbitration system and the immune system, have proposed a new decentralized consensus-making system inspired by the biological immune system for engineering application [48–50, 59, 89].

This section introduces immune algorithm–based schemes into control engineering area with the goal of effectively controlling nonlinear systems, complex systems with disturbances, including their tuning.

2.2 Immune System for Intelligent Parameter Estimation

The immune system has two types of response: primary and secondary. The primary response is a reaction when the immune system encounters the antigen for the first time. At this point, the immune system learns about the antigen, thus preparing the body for any further invasion from that antigen. This learning mechanism creates a memory function in the immune system. The secondary response occurs when the same antigen is encountered again. This has response characterized by a more rapid and more abundant production of antibodies resulting from the priming of the B-cells (B-lymphocytes) in the primary response.

When a naive B-cell encounters an antigen molecule through its receptor, the cell gets activated and begins to divide rapidly; the progeny derived from these B-cells differentiate into memory B-cells and effector B-cells or plasma cells. The memory of B-cells has a long life span and they continue to express membrane bound antibodies with the same specificity as the origin parent B-cell [43, 46, 70].

Both neural networks and immunity-based systems are biologically inspired techniques that have the capability of identifying patterns of interest. They use learning, memory, and associative retrieval to solve recognition and classification tasks. But the underlying mechanisms of recognition and learning are very different. The immune system provides diversification instead of converging to local or global optima. The immune system possesses self-organizing memory and remembers its categorizations over long periods of time [46].

However, artificial neural networks are similar structures that learn to associate patterns of input data with an appropriate output response by modifying their connectivity according to a defined learning algorithm. Specifically, neural networks are concerned with approximating input–output relationships, with many neurons linked together by connections of varying strength. The actual manner in which these connections are made determines the flow of information in the network and defines the network model. There exist many different immune network models as defined in other materials. With distinct performance features [43], they have been interesting for widely real-world applications such as signal processing, speech recognition, intelligent control, robotics, and others for distribution system.

From an information-processing perspective, the immune system is a distributed system. It provides an excellent model of adaptive processes operating at the local level or at the global level. F. J. Varela [48] has compared the immune system and the nervous system. He has shown many similarities between the two systems at the level of system behavior, though they differ at the respective building-block level. He postulated a symmetrical neural network model that can produce desired stimulus-response behavior similar to the immune response. Other researchers have also drawn some analogies between the two systems [43, 49]. The biological metaphor behind the artificial neural networks is that the brain neurons are connected via synapses where the neurons receive signals from other neurons or from external stimulii.

The artificial immune system is also a very complex system with several defense mechanisms in our body. There are various cell types that can attack invaders directly, and they can also secrete molecules with a variety of functions, including attacking foreign cells and signaling other immune cells to proliferate. Lymphocytes are the most important class of cells. Unlike neurons, lymphocytes float freely in blood. In particular, they patrol everywhere for foreign antigens, then gradually drift back into the lymphatic system to begin the cycle all over again. Each lymph node contains specialized compartments where they can also encounter antigens. They interact with antigens at a rate dependent upon the concentration and affinity of the antigens. There are two main types of lymphocytes, namely T cells and B cells, each having receptors on its surface to respond to a limited group of structurally related antigens.

Self–non-self discrimination is one of the main tasks the immune system solves, and negative selection is one of the principle mechanisms in the chain of events that result in robust self–non-self discrimination. This discrimination is achieved in part by T cells. This process eliminates the self-reacting T cells. Specifically, the random generation of variation of cells followed by negative selection creates circulating T cells that react only with non-self molecules. The processes by which B cells are prevented from reacting to self molecules are not fully understood [53, 55].

It may be that the absence of self-reactive T cells is sufficient to prevent clonal expansion of self-reactive B cells. Some T cells called, the killer T cells, destroy an infected self cell if they recognize an infection from external condition. Others are helper T cells which trigger clonal expansion of B cells. Affinity maturation occurs when the mutation rate of a B cell clone increases in response to a match between the clone's antibody and an antigen. Variable cell division and programmed cell death rates allow the immune system to allocate resources, such as cells, dynamically and when needed in a distributed environment. Although the models of neural networks and their applications have received wide attention, immune system models are of limited use so far. However, both neural system and immune system share many common features as basic cognitive mechanisms.

In the immune system, learning occurs through the modification of the number and affinities of the receptors. However, connections within a nervous system are deterministic, whereas the cross reaction in the immune response is a stochastic process.

2.3 Dynamic Models of Artificial Immune System for Parameter Estimation

2.3.1 Overview for Parameter Estimation

There are several models to study both general properties of immune systems and specific immunological phenomena [51–59]. The goal of these models is to deduce macroscopic principles of the system from the properties and interactions among the components of the immune system. Moreover, the natural immune system is also a source of inspiration for developing intelligent problem solving methodologies. A survey of models that have been developed based on various computational aspects of the immune system and their applications to real world problems are available through these reasons [59].

Forrest et al. [53] developed a negative-selection algorithm for change detection based on the principles of self or non-self-discrimination by T cells [15]. In this approach, the self is defined as the normal pattern of activity of a system or a collection that they wish to protect or monitor against the external conditions. A diverse set of detectors is generated in the complement space of the self, similar to the censoring process of mature T cells. Then the detectors are used to monitor self for changes by continually matching the detectors against the representative of self. If any detector ever matches, a change must have occurred in self. This algorithm exhibits many interesting properties in change detection. Dasgupta and Forrest experimented with several data sets to investigate the performance of the negative selection algorithm for detecting anomaly in the data series. In these experiments, the notion of self is considered as the normal behavior patterns of the monitored system. In particular, the abnormal detection problem is reduced to the problem of detecting whether or not an observed data pattern has changed, where match implies a shift in the normal behavior patterns. Dasgupta and Forrest observed the following similarities and differences between the negative selection algorithm and the ART network in their experiments on anomaly detection in time series data [47].

Like the nervous system, the immune system can learn new information, recall previously learned information, and perform pattern recognition tasks in a highly decentralized system. In particular, it learns to recognize relevant patterns, remember patterns that have been seen previously, and use combinatorics to construct pattern detectors efficiently. Also, the overall behavior of the immune system is an emergent property of many local interactions [56]. It has been suggested that the immune system functions as a kind of second brain because it can store memories of past experiences in strengths of the interactions of its constituent cells, and it can generate responses to novel patterns such as new antigens. Furthermore, the immune response develops in time and the description of its time evolution is an interesting problem in the framework of the dynamical systems theory [50]. Similar to the artificial neural network, the artificial immune system is also trained to learn the dynamics and the statistical properties of the monitored system. Like in the artificial neural network algorithms, it is necessary to tune control parameters of artificial immune system to get the best performance. The associative memory characteristics of the Hopfield network are also present in the immunological models [54].

Many research results explain that the immune system is in principle more complex than the brain-nervous system. Different theories have been suggested based on experimental evidences. Some of these theories are contradictory regarding the way cellular interactions occur in the immunological response. Because of the lack of understanding, there is still no unified model or view that would account for all known immunological phenomena. Also there remain open questions whether the immune system operates at a steady state, oscillates or is chaotic. However, the remarkable pattern recognition abilities of animal immune systems suggest that immune system-based algorithms are well worth exploring. Future research should investigate various computational properties of the immune system, improvements to the existing models, and new application areas.

2.3.2 Antibodies of Immune System in Parameter Estimation

The antibody molecule acts as a bridge between the target cell, subsequently causing the target cell due to activation of cytotoxic cell through receptor. An antibody is actually a three-dimensional Y-shaped molecule which consists of two types of protein chains: light and heavy. It also possesses two paratopes which represent the pattern it will use to match the antigen. The regions on the molecules that the paratopes can attach are the so-called epitopes. These same molecules with antigenic peptide bound to them will be responsible for interaction with the T-cell receptor. The site on an antigenic peptide that interacts with a T-cell receptor is called an epitode [43, 56].

2.3.3 Interaction between Antibodies in Parameter Estimation

The antigen antibody interaction is similar to that of enzyme substrate interaction except that this interaction does not lead to irreversible alter-action in either antibody or antigen and therefore is reversible. The reaction between an antigen and antibody is of noncovalent type, where the antigenic determinants or epitodes interact with the domain of the antibody molecule. The noncovalent interaction between antigen and antibody is brought about by hydrogen bonds, van der Waals interactions, ionic bonds and hydrophobic interactions. Therefore, a strong affinity interaction should occur between antigen and antibody to form a stable complex [42, 49, 50]. In Fig. 2.1, figuring out the interaction among antibodies is important to understand the dynamic characteristics of immune system. These antigens stimulate the antibodies, consequently the concentration of antibodies A1 and A2 increases. However, if there is no interaction between antibody A1 and antibody A2, these antibodies will have the same concentrations. Suppose that the idiotope of antibody A1 and the paratope of antibody A2 are the same. This means that antibody A2 is stimulated by antibody A1, and oppositely antibody A1 is suppressed by antibody A2, as shown Fig. 2.1. In this case, unlike the previous case, antibody A2 will have higher concentration than antibody A1. As a result, antibody A2 is more likely to be selected. This means that antibody A2 has higher priority over antibody A1 in this situation [56, 58, 59].

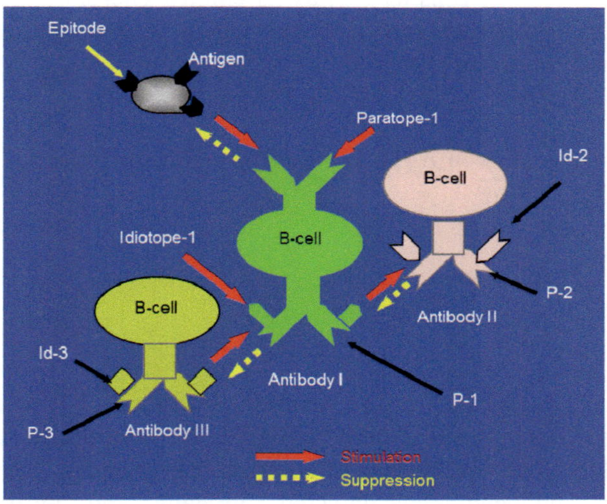

Figure 2.1(a) Stimulation and suppression action between antibody and antigen on an immune network.

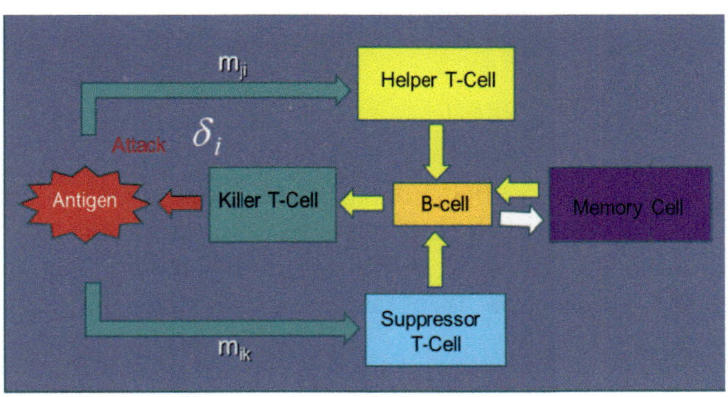

Figure 2.1(b) Dynamic relationship between cells, antigen.

2.3.4 Dynamic Equation of Immune System for Parameter Estimation

As mentioned above, in the immune system, slightly different dynamical model can be suggested depending on research. However, as shown in Fig. 2.1, basically, the level to which a B cell is stimulated relates partly to how well its antibody binds the antigen. This section takes into account both the strength of the match between the antibody and the antigen and the B cell's affinity to other B cells as well as its enmity. Therefore, generally the concentration of i-th antibody, which is denoted by δ_i, is calculated as follows [42–50]:

$$\frac{dS_i(t)}{dt} = \left(\alpha \sum_{j=1}^{N} m_{ji}\delta_j(t) - \alpha \sum_{k=1}^{N} m_{ik}\delta_k(t) + \beta m_i - \gamma_i\right)\delta_i(t) \tag{2.1a}$$

$$\frac{d\delta_i(t)}{dt} = \frac{1}{1+\exp\left(0.5 - \dfrac{dS_i(t)}{dt}\right)}, \tag{2.1b}$$

where in Eq. (2.1a), N is the number of antibodies, α and β are positive constants. m_{ji} denotes affinities between antibody j and antibody i (i.e., the degree of interaction), finally m_i represent affinities between the detected antigens and antibody i, respectively.

On the other hand, information obtained in the lymphocyte population can be represented by [17]:

$$\Omega_j(N) = \sum_{i=1}^{S}(-x_{ij}\log x_{ij}), \tag{2.2}$$

where N is the size of the antibodies in a lymphocyte population, S is the variety of allele and x_{ij} represents the probability that locus j is allele i. Therefore, the means of information $\Omega_{ave}(N)$ in a lymphocyte population is obtained as the following equation [17, 39]:

$$\Omega_{\text{ave}}(N) = \frac{1}{M}\sum_{j=1}^{M}\Omega_j(N)$$
$$= \frac{1}{M}\sum_{j=1}^{M}\left\{\sum_{i=1}^{S} -x_{ij}\log x_{ij}\right\}, \qquad (2.3)$$

where M is the size of the gene in an antibody.

The affinities m_{ji} between antibody j and antibody i are calculated as follows:

$$m_{ji} = \frac{1}{\{1+\Omega(ji)\}} \qquad (2.4a)$$

$$\Omega(ji) = H_s(x) = \left[f_1(x) + f_2(x) + f_3(x)\right] \qquad (2.4b)$$

where $\Omega(ji)$ is the information obtained by antibodies j and i.

If $\Omega(ji) = 0$, the antibody j and antibody i match completely. Generally m_{ji} is given in range of [0, 1]. In Eq. (2.4b), $f_1(x)$, $f_2(x)$ and $f_3(x)$ means objective function of control system that we are going to design.

2.4 Optimal Nonlinear Parameter Estimation of Induction Motor Using Immune Algorithm

2.4.1 Background

In an AC induction motor drive, the electrical parameters are in general determined via the classical analysis and no-load tests. Estimation of the performance behavior of an induction machine is also done by plotting the steady-state slip curves. Generally to obtain the parameter values, one must use the equivalent circuit relations and the experimental results obtained from the above-mentioned classical analysis. Therefore, the parameter values obtained by direct classical approaches or experimentation can reveal significant differences in the entire range of slip varying from 0 to 1. To describe the performance of the induction machine more precisely and to reduce the differences between the estimated and real performances, one must modify the parameters obtained from the classical analysis. To achieve this purpose in motor control, the use of system identification algorithms based on the artificial immune algorithm appears to be a very promising approach. These algorithms allow one to take into account the effect of measurement errors, disturbances, and random signals on the estimated parameters. Since the equations relating the phase current to the slip and the circuit parameters involve many variables and are nonlinear, parameters can have difference values in case of the change of load.

This fact does not enable one to directly use the many parameter-estimating procedures from in the literature.

In this section, obtaining optimal parameters of the equivalent circuit of a squirrel-cage induction machine by an immune algorithm is suggested. During the execution of the estimation algorithm, we use the steady-state characteristic curves of both the input power and the stator current to adjust the initial parameter vector. The difference between the proposed immune-based optimal parameters and those based on the classical procedure are analyzed.

2.4.2 Induction Motor Model for Parameter Estimation

A squirrel-cage induction machine supplied with a three-phase symmetrical voltage source can be described using the equivalent circuit shown in Fig. 2.2.

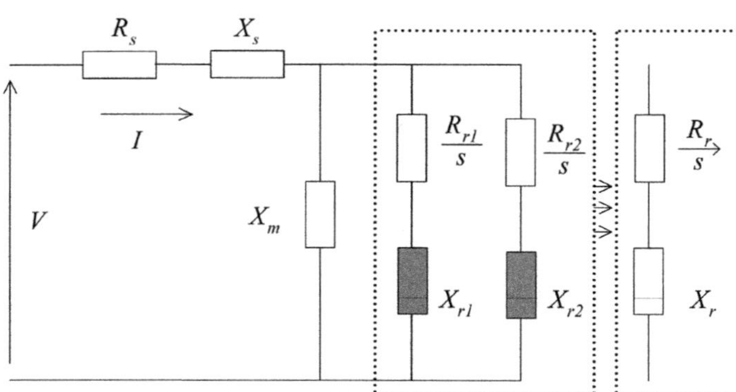

Figure 2.2 Equivalent circuit of a squirrel-double induction motor.

In case the stator current, the input power, the equations, and the electromagnetic torque for a squirrel induction motor can be deduced from the circuit of Fig. 2.2 and are expressed as follows:

$$I(s) = V\sqrt{\frac{C^2 + D^2}{A^2 + B^2}}$$

$$P(s) = 3V^2 \frac{AC - BD}{A^2 + B^2}$$

$$T(s) = 3V^2 \frac{p}{\omega} \frac{R_{fe}^2 R_r / s}{A^2}$$

$$A = R_s \left(1 + \frac{X_r}{X_m}\right) + \left(1 + \frac{X_s}{X_m}\right)\frac{R_s}{S}$$

$$B = X_r + X_s \left(1 + \frac{X_r}{X_m}\right) - R_s \left(R_r / S \ \frac{R_r / S}{X_m}\right) \tag{2.5}$$

$$C = 1 + \frac{X_r}{X_m}$$

$$D = \frac{R_r / S}{X_m}$$

$$R_r = \frac{R_{r1}R_{r2}(R_{r1} + R_{r2}) + (R_{r1}X_{r2}^2 + R_{r2}X_{r1}^2)S^2}{(R_{r1} + R_{r2})^2 + (X_{r1} + X_{r2})^2 S^2}$$

$$R_r = \frac{X_{r1}X_{r2}(X_{r1} + X_{r2})S^2 + R_{r1}^2 X_{r2} + R_{r2}^2 X_{r1}}{(R_{r1} + R_{r2})^2 + (X_{r1} + X_{r2})^2 S^2}$$

$$I(s) = V\sqrt{\frac{C^2 + D^2}{A^2 + B^2}}$$

$$P(s) = 3V^2 \frac{AC - BD}{A^2 + B^2}$$

In the above equation, R_s, R_r, and are the stator, rotor, and iron losses, respectively. Also, X_s X_r and X_m are the stator leakage reactance, rotor leakage reactance, and magnetizing reactance. For neglecting the iron losses of a double-cage motor, one must add a second branch in parallel with the magnetizing reactance.

2.4.3 Optimal Parameter Selection for Parameter Estimation

In order to determine model parameters from the slip curves of the equivalent circuit, references [123] and [124] use a nonlinear curve-fitting problem stated as the solution to the following minimization problem:

$$\min_{\phi \in \Gamma} = J(\phi) = \frac{1}{N}\sum_{i=1}^{N}\left[y_i - y(s_i, \phi)\right]^2, \tag{2.6}$$

where $J(\phi)$ is least-squares cost function obtained by the sum of the squares of the differences between the experimental and calculated slip curves, Γ is parameter space depending on the number of parameters to be estimated, y_i is the experimental data value collected from machine, $y(s_i, \phi)^2$ is a nonlinear function relating the measured data, the circuit parameters, and the slip, and ϕ is a parameter vector pertaining to Γ. Therefore, in case of a double cage, the dimension of parameter vector is defined as:

$$\phi = \left[R_{ri}, R_{r2}, X_{r1}, X_{r2}, X_s, R_s, X_m\right]^T. \tag{2.7}$$

This equation depends on the kind of available experimental data and obtaining a parameter vector that minimizes the quadratic performance index defined by Eq. (2.6). In this case, since one must deal with a nonlinear algorithm to acquire the desired solution, some numerical problems may arise or a direct approach would require writing down the normal equations for solving them. The methods for numerical minimization of performance index Eq. (2.7) might be modified to update the estimated parameter vector according to load change.

2.4.4 Clonal Selection Algorithm for Optimal Parameter Selection

In this chapter, clonal selection algorithm in introduced as depicted in Fig. 2.1 of the previous section. That is, when an antibody on the surface of a B cell binds an antigen, that B cell becomes stimulated. The level of stimulation depends on the B cell not only on how well the B cell's antibody matches the antigen but also how it matches other B cells in the immune network [53, 54, 59], called its affinity with other B cells. The immune network is formed by B cells.

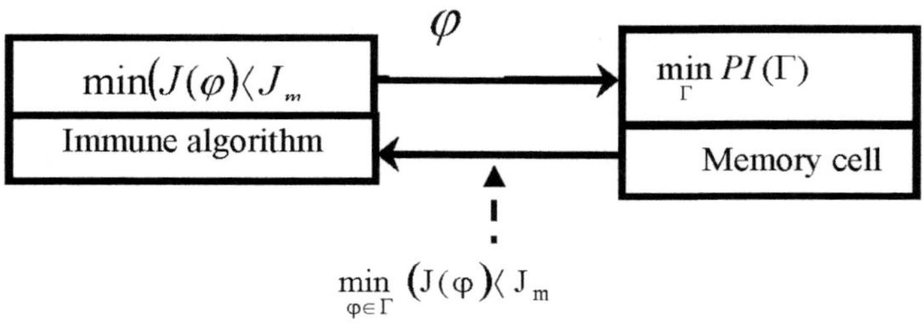

Figure 2.3 Immune algorithm-based computational structure for optimal parameter estimation.

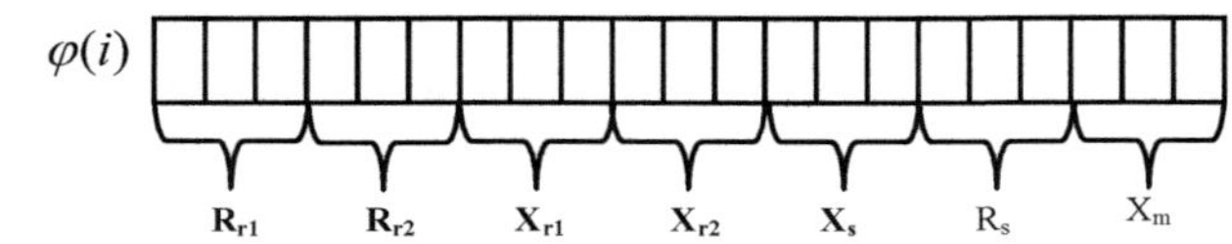

Figure 2.4 Allocation structure of each parameter.

If the stimulation level rises above a given threshold, the B cell becomes enlarged and if the stimulation level falls below a given threshold, the B cell dies off.

The more neighbors the B cell has an affinity with, the more stimulation it will receive from the network, and vice versa. The level to which a B cell is stimulated relates partly to how well its antibody binds antigen.

We take into account both the strength of the match between the antibody and the antigen and the B cell object's affinity to other B cells as well as its enmity. Therefore, generally the concentration of *i*-th antibody, which is denoted by δ_i, is calculated as in Eq. (2.1).

The coding of an antibody in the immune network is very important because a well designed antibody coding can increase the efficiency of the controller. The efficiency of estimation will be performed in this section. As shown in Fig. 2.2, there are seven antibodies for parameters, $\phi = [R_{ri}, R_{r2}, X_{r1}, X_{r2}, X_s, R_s, X_m]^T$

and an object function for Eq. (2.7). Each parameter is specified in the memory cell of the immune network.

2.4.5 Computational Procedure for Optimal Estimation

[Step 1] Initialization and recognition of the antigen: The immune system recognizes the invasion of the antigen, which corresponds to the parameter for the optimization problem as shown in Fig. 2.2.

[Step 2] Production of antibodies from the memory cell: The immune system produces the antibodies that were effective to kill the antigen in the past. This is implemented by recalling a past successful solution from the memory cell.

[Step 3] The antibody with the best fitness value obtained by calculation for searching an optimal solution is stored in the memory cell.

[Step 4] Differentiation of lymphocytes: The B-lymphocyte cell, the antibody that matched the antigen, is dispersed to the memory cells in order to respond to the next invasion quickly. That is, select the individuals using tournament selection approach and apply genetic operators (crossover and mutation) to the individuals of network.

[Step 5] Stimulation and suppression of antibody: The expected value of the stimulation of the antibody is given by

$$\eta_k = \frac{m_{\phi k}}{\sigma_k}, \tag{2.8}$$

where σ_k is the concentration of the antibodies. The concentration is calculated by affinity. So, σ_k is represented by

$$\sigma = \frac{\text{sum of antibodies with same affinity as } m_{\phi k}}{\text{sum of antibodies}} \tag{2.9}$$

Using Eq. (2.9), an immune system can control the concentration and the variety of antibodies in the lymphocyte population. If the antibody obtains a higher affinity against an antigen, the antibody stimulates. However, an excessive higher concentration of an antibody is suppressed. Through this function, an immune system can maintain the diversity of searching directions and a local minimum.

[Step 6] Calculate fitness value between antibody and antigen. This procedure can generate a diversity of antibodies by a genetic reproduction operator such as mutation or crossover. These genetic operators are expected to be more efficient than the generation of antibodies. Table 2.1 shows the results obtained by clonal selection.

[Step 7] If the maximum number of generations of memory cell is reached, stop and return the fitness of the best individual fitness value in the network; otherwise, go to step 3.

Table 2.1 Initial boundary and true values

gen	R_{r1}	R_{r2}	X_{r1}	X_{r2}	X_s	R_s	X_m
ϕ_{ture}	0.0693	0.0132	0.00843	0.1162	0.123	0.00778	4.3
xlb	0.06	0.01	0.007	0.10	0.10	0.006	4
xub	0.08	0.015	0.01	0.13	0.13	0.008	4.5

2.5 Simulation and Discussion

The clonal selection algorithm for parameter estimation suggested in this section is simulated and compared with genetic algorithm and the recursive algorithm cited in reference [123]. In reference [123], objective function $J_1(\varphi)$ is used but object function $J_2(\varphi)$ is introduced for more optimal parameter selection as follows:

$$J_1(\varphi)=\frac{1}{N}\sum_{i=1}^{N}[I(si)-I(si,\varphi)]^2+\frac{1}{N}\sum_{i=1}^{N}[P(si)-P(si,\varphi)]^2 \tag{2.10}$$

$$J_2(\varphi)\frac{1}{N}\sum_{i=1}^{N}[S(i)I(si)-S(i)I(si,\varphi)]^2+\frac{1}{N}\sum_{i=1}^{N}[S(i)P(si)-S(i)P(si,\varphi)]^2 \tag{2.11}$$

Figure 2.5 represents a variation of current curve I(s) to the slip of induction motor obtained by the clonal selection algorithm combined with genetic algorithm (CS-GA) and true values.

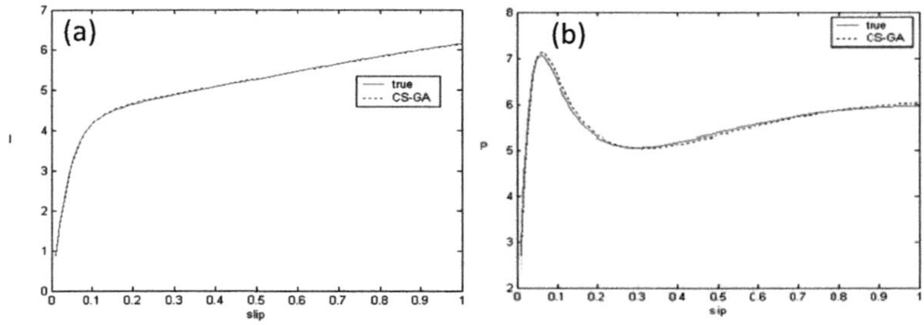

Figure 2.5 Variation of I (a) and P (b) by the clonal selection.

Figure 2.6 is variation of values of I (a) and P (b) obtained by clonal selection method genetic algorithm, recursive algorithm, and the true values. Figure 2.7 shows parameter variation to population in objective function $J_1(\varphi)$ and $J_2(\varphi)$ as function of slip for various populations.

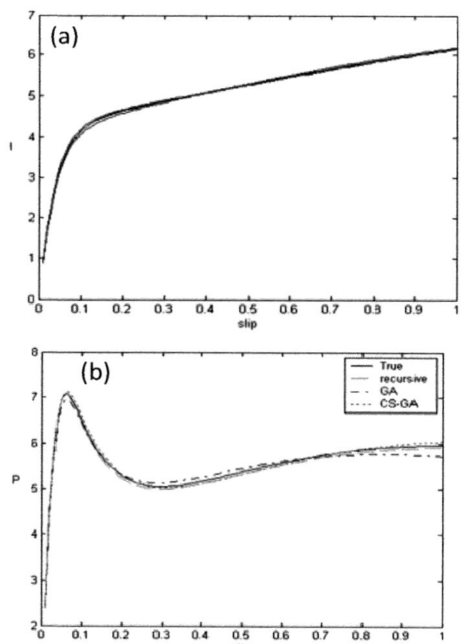

Figure 2.6 Variation of values of I (a) and P (b) obtained by clonal selection method genetic algorithm, recursive algorithm, and the true values.

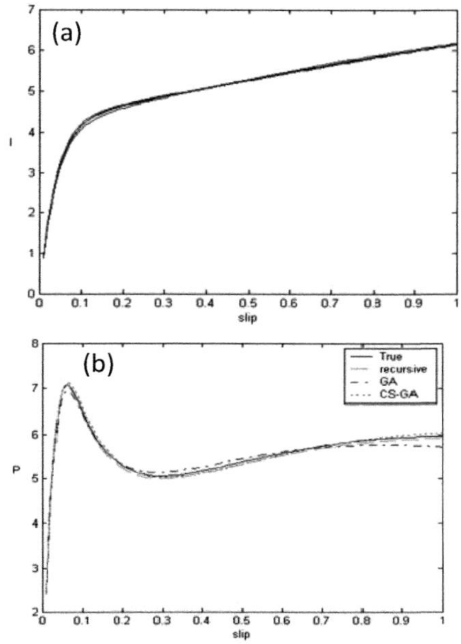

Figure 2.7 Parameter variation to population in objective function $J_1(\varphi)$ and $J_2(\varphi)$ as function of slip for various populations.

Figure 2.8 represents comparison of two objective function ($J_1(\varphi)$ and $J_2(\varphi)$) parameters to generations of clonal selection in each objective function.

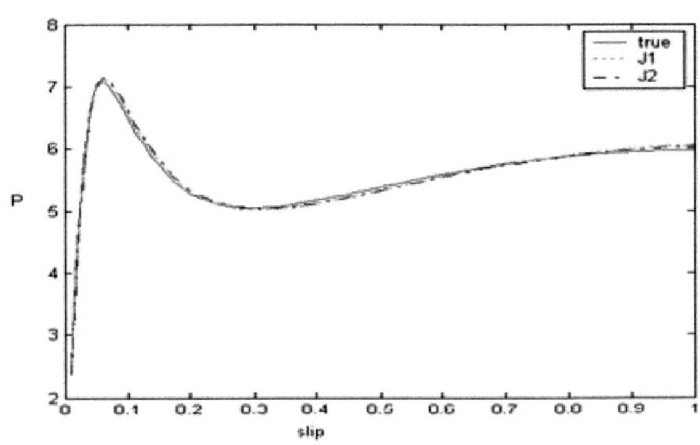

Figure 2.8 Comparison of two objective function ($J_1(\varphi)$ and $J_2(\varphi)$) parameters to generations of clonal selection in each objective function.

Table 2.1 is the initial boundary of parameters for estimation and true values and Table 2.2 gives parameters estimated using GA (Genetic Algorithm), and recursive method, clonal selection suggested in this section. Table 2.3 is comparison of parameters to popsize of clonal selection in each objective function. That is, Table 2.3 shows parameters of each objective function and simulated parameters in motor parameters.

Table 2.2 Parameter estimated by immune algorithm

Item	R_{r1}	R_{r2}	X_{r1}	X_{r2}	X_s	R_s	X_m
Recursive	0.078	0.0129	0.0164	0.121	0.1167	0.0073	4.29
GA	0.063	0.0138	0.010	0.122	0.1260	0.0074	4.21
CS-GA	0.0755	0.0135	0.009	0.1168	0.1195	0.0064	4.34

Table 2.3 Parameters of each objective function and simulated parameters

Obj function	Popsize	R_{r1}	R_{r2}	X_{r1}	X_{r2}	X_s	R_s	X_m
	60	0.0755	0.0135	0.0089	0.1168	0.1195	0.0064	4.34
$(J_1(\phi))$	100	0.0647	0.0132	0.0081	0.1186	0.1253	0.0078	4.32
	150	0.0699	0.0115	0.0074	0.1231	0.1207	0.0066	4.42
	60	0.0755	0.0135	0.0091	0.1159	0.1196	0.0068	4.28
$(J_2(\phi))$	100	0.068	0.0149	0.0081	0.1174	0.1242	0.0069	4.47
	150	0.0787	0.0136	0.0082	0.1105	0.1188	0.0077	4.30

2.6 Summary

The use of various identification techniques in determining the values of the equivalent circuit parameters of a three-phase squirrel-cage induction machine has been suggested. In traditional approaches such as the recursive method, the algorithms have the problem of dealing with ill-conditioned matrices. Such matrices usually occur in nonlinear problems because identification procedures are based on the steady-state phase current versus slip and input power versus slip characteristics. Traditional approaches may not able to give satisfactory results for nonlinear systems since the traditional identification algorithms are also of a linear kind because machine parameters are obtained as solutions to the least-squares cost function minimization of the difference between calculated and experimental steady-state characteristics.

Table 2.4 Parameters by generation of the each objective function

Obj function	gen	R_{r1}	R_{r2}	X_{r1}	X_{r2}	X_s	R_s	X_m
$(J_1(\varphi))$	100	0.075594	0.013538	0.0089178	0.11685	0.11951	0.0064439	4.3419
	150	0.072702	0.01227	0.007611	0.11916	0.1202	0.0069193	4.1168
	200	0.072702	0.01227	0.007611	0.11916	0.1202	0.0069193	4.1168
	300	0.072702	0.01227	0.007611	0.11916	0.1202	0.0069193	4.1168
$(J_2(\varphi))$	100	0.075521	0.013558	0.0091887	0.11596	0.11962	0.0068187	4.2845
	150	0.075521	0.013558	0.0091887	0.11596	0.11962	0.0068187	4.2845
	200	0.075495	0.013516	0.0092333	0.11596	0.11961	0.0068168	4.4295
	300	0.0755	0.01356	0.0092373	0.11596	0.11963	0.0068197	4.3595
	400	0.0755	0.01356	0.0092379	0.11596	0.11963	0.0068197	4.3604

In this section, nonlinear parameter estimation approach is suggested for a 1.5 kW wound-rotor three-phase induction motor, using the clonal selection in immune algorithm. The estimation algorithm method yields good parameter estimates, without the necessity to constrain the parametric vector. The algorithm converges quickly (around 100 generations in Table 2.3). It is very simple to implement and can use data collected in a wide range. The algorithm enables estimation without increasing the computational complexity. The presented approach allows one to deal with a parameter vector that includes variables carrying information on a nonlinear system, such as the varying load in induction motor.

Chapter 3

Intelligent System Tuning Using a Hybrid GA-PSO Approach

3.1 Background

When individuals in GA are differentiated to search for optimal solutions, there is a high chance for obtaining local optimal solutions. However, it can be easily localized sometimes. Particle swarm optimization (PSO) conducts searches using a population of particles that correspond to individuals like GA. Initially a population of particles is randomly generated and each particle represents a potential solution, and has a position represented by a position vector [14, 117]. A swarm of particles moves through the problem space, with the moving velocity of each particle represented by a velocity vector. At each time step, a function representing a quality measure is calculated by using Euclidean data distance as input in control system. Each particle keeps track of its own best position, which is associated with the best fitness it has achieved so far in a vector. Furthermore, the best position among all the particles obtained so far in the population is kept track as output. In addition to this global version, another local version of PSO keeps track of the best position among all the topological neighbors of a particle. At each time step, by using the individual best position, and global best position, a new velocity for particle is updated using

Tuning Innovation with Biotechnology
Dong Hwa Kim

ISBN 978-981-4745-35-2 (Hardcover), 978-1-315-36458-2 (eBook)
www.panstanford.com

Eq. (3.1) in this paper. In Eq. (3.1), uniformly distributed random numbers in [0, 1] are used as positive constants. If the velocity violates this limit, it is set at its proper limit. Changing velocity this way enables the particle to search around its individual best position, and global best position [106, 107] for optimal solution. The computation of PSO is easy and adds only a slight computational load when it is incorporated into GA. Furthermore, the flexibility of PSO to control the balance between local and global exploration of the problem space helps overcome premature convergence of elite strategy in GA and also enhances search ability. In the following section, the detailed algorithm for the hybrid of PSO with GA is introduced.

This section focuses on the advantages of introducing PSO into the mutation process of GA, for improving the GA learning efficiency. Euclidean distance is used in the crossover to avoid local optima and to obtain fast convergence of solution. We illustrate the performance of the method using four test functions.

3.2 Euclidean Data Distance and Optimal Solution

Using the conventional GA or PSO approach optimal solutions are obtained mostly with some initially the differentiated data and there is a high possibility for obtaining local optimal solutions. The proposed approach uses data points with the longest Euclidean distance for crossover process to avoid such local optimization. The idea is to obtain global solutions by considering the entire search space (all the data points). Figure 3.1 illustrates the concept of Euclidean distance and Fig. 3.2 represents the max and min of the Euclidean distance. Euclidean distance for the function

$$F_1(x) = \sum_{i=1}^{3} x_i^2, \tag{3.1}$$

and thus a local solution could be avoided. The distance between two points on n search space is defined by (Fig. 3.1)

$$\text{Distance} = \sqrt{(x_1 - y_2)^2 + (x_2 - y_2)^2 + \cdots + (x_n - y_n)^2} \tag{3.2}$$

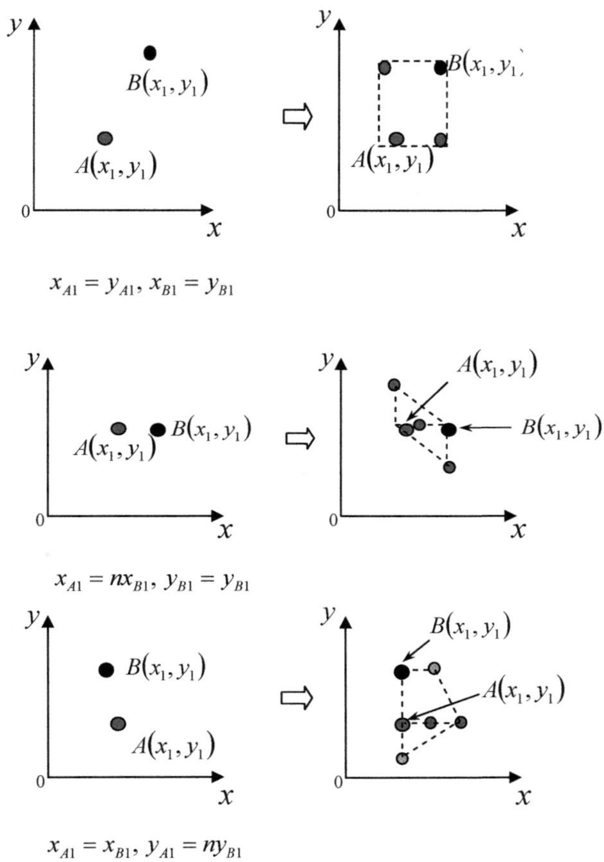

Figure 3.1 Definition of Euclidean data distance for GA.

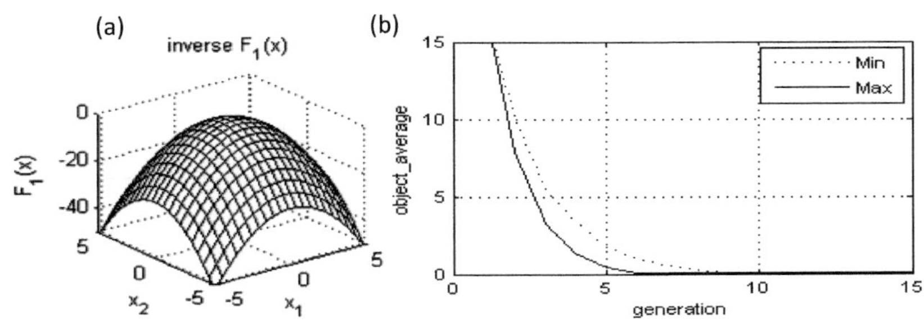

Figure 3.2 Max and Min of the Euclidean data distance (a) and the value of the average objective function (b) with respect to the number of generations for the test function F_1 using GA.

To demonstrate the performance, the Himmelblau function is used:

$$F(x) = (x^2 + x_2 - 11)^2 + (x_1 + x_2^2 - 7)^2 \quad (3.3)$$

The contour to obtain optimal solution by crossover using a conventional GA is illustrated in Fig. 3.1. Data points are selected by

$$A(x_1, y_1) \oplus B(x_1, y_1) \Rightarrow A', B' \left(x_1' \Big|_{\min(x_1, y_1)}^{\max(x_1, y_1)}, y_1' \Big|_{\min(x_1, y_1)}^{\min(x_1, y_1)} \right) \quad (3.4)$$

As per the contour plot (Fig. 3.1) it is evident that there is an optimal solution in only one place and optimal solution is obtained after 200 generations. Figure 3.2 shows that Max and Min of the Euclidean data distance and the value of the average objective function to the number of generation of GA. The contour processing plot (Fig. 3.3) obtained by the proposed GA based on Euclidean distance shows that there are optimal solutions in both local and global spaces and its solution is obtained after 50 generations. The influence of mutation in GA or in a hybrid system of PSO and GA has been studied to speed up the running time to obtain optimal solution [3–5].

Table 3.1 is the initial condition of test function $F(x)_1 = \sum_{i=1}^{3} x_i^2$ to prove suggested algorithm in this session and Table 3.2 shows results optimal value of objective function and average value of objective function through simulation of test function $F(x)_1 = \sum_{i=1}^{3} x_i^2$.

Table 3.1 Initial condition for performance

	Definition			
Function	$x_i^{(L)}$	$x_i^{(U)}$	**ID**	**IT**
$F(x)_1 = \sum_{i=1}^{3} x_i^2$	5.12	5.11	60	100

Table 3.2 Result by min-max method

	x_1	x_2	Optimal value of objective function	Average value of objective function
Max	1.0885e-009	7.1709e-010	1.6991e-018	3.5601e-013
Min	−2.2190e-011	1.0253e-009	1.0518e-018	3.7901e-013

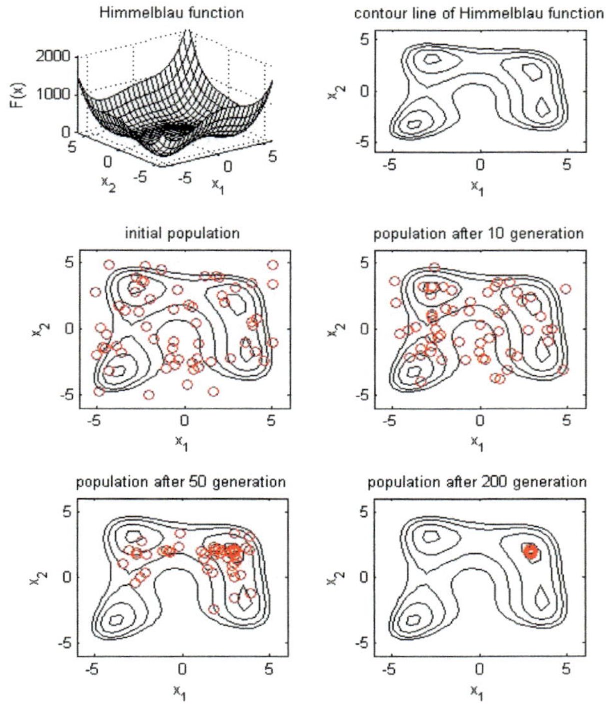

Figure 3.3 Contour of optimal procedure by GA.

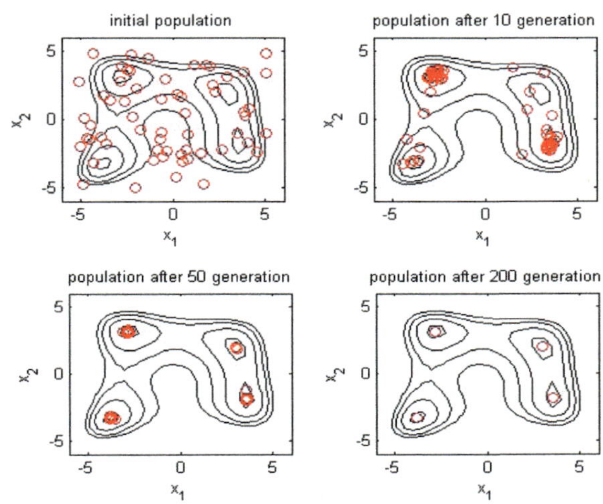

Figure 3.4 Himmelblau function: contour of GA approach incorporating the Euclidean distance.

3.3 Particle Swarm Optimization

3.3.1 Overview

This section describes the PSO concept in terms of its precursors, briefly reviewing the stages of its development from social simulation to optimizer for science and engineering area. The implementation of one paradigm will be discussed more in the next section, followed by results obtained from applications and test upon which the paradigm has been shown to perform successfully in system, especially control engineering. PSO has approaches in two main component methodologies. It can be found in general such as, bird flocking, fish schooling, ants, and so on by Reynolds [8] and Heppner and Grenander [4].

It is also related, however, to evolutionary computation and has ties to both genetic algorithms and evolutionary programming [106–108]. PSO developed by the concept, and paradigms can be implemented in a few lines of computer code. It requires only primitive mathematical operators and is computationally inexpensive in terms of both memory requirements and speed for optimal solution. Early testing has found the implementation to be effective with several kinds of problems. This section suggests and discusses application of this algorithm to the training of algorithms.

A number of researchers have been exploring computer simulations of various interpretations of the movement of natural from such as, a bird flock or fish school. Notably, Reynolds [8], Heppner, and Grenander [4] presented simulations of bird flocking. Reynolds was intrigued by the aesthetics of bird flocking choreography, and Heppner, a zoologist, was interested in discovering often changing direction suddenly, scattering and regrouping, etc. These models relied heavily on the manipulation of inter-individual distances; that is, the synchrony of flocking behavior was thought to be a function to maintain an optimum distance between themselves and their neighbors. In the case of fish schoolings, E. O. Wilson [9] has suggested that at least individual members of the school can profit from the discoveries and previous experience of all other members of the school during the search for food. This advantage become decisive, whenever the resource is unpredictably distributed in patches [105–107].

These results mentioned that social sharing of information among offering an evolutionary advantage: this hypothesis was fundamental to the development of particle swarm optimization. Birds and fish adjust their physical movement to avoid predators, seek food and mates, optimize environmental parameters such as temperature, etc. Humans adjust not only physical movement but cognitive or experiential variables as well. PSO has been shown to successfully optimize a wide range of continuous functions [105–107] in engineering area. The individual particles are driven stochastically toward the positions of their own previous best performance or their neighbors. A generalized model is proposed, including methods for controlling the convergence properties of the particle system. Some empirical results are given on the performance of various implementations in engineering area.

PSO is an extremely simple algorithm that seems to be effective for optimizing a wide range of functions. The adjustment toward pbest and gbest of the particle swarm optimizer is conceptually similar to the crossover operation utilized by genetic algorithms. It uses the concept of fitness in evolutionary computation.

The goals in developing in this paper keep it simple and robust, and global optimization. The algorithm is written in a very few lines of code, and requires only specification of the problem and a few parameters in order to solve it. Once again nature has been providing us with a best technique for processing information and idea that is creative at once versatile.

3.3.2 PSO-Based Optimization

A population of particles is initialized with random positions $\vec{k}_i$ and velocities $\vec{v}_i$ and a function, f, is evaluated, using the particle's positional coordinates as input values. Positions and velocities are adjusted, and the function evaluated with the new coordinates at each time-step. When a particle discovers a pattern that is better than any it has found previously, it stores the coordinates in a vector, pbest$_j$. The difference between pbest$_j$ (the best point found by i so far) and the individual's current position gbest is stochastically added to the current velocity, causing the trajectory to oscillate around that point. Further, each particle is defined within the context of a topological neighborhood comprising itself and some other particles in the population. The stochastically weighted difference between the neighborhood's best position gbest and the individual's current position is also added to its velocity, adjusting it for the next time-step. These adjustments to the particle's movement through the space cause it to search around the two best positions.

We used the position and velocity vector of PSO as follows [105–107]:

$$v_{f,g}^{(t+1)} = w \cdot v_j^{(t)} + c_1^* \text{ Rand}(\text{pbest}_{j,g} - k_{j,g}^{(t)})^* + c_2^* \text{ Rand}(\text{gbest}_g - k_{j,g}^{(t)})^* \tag{3.5}$$

where n is the number of agent in each group, m the number of member in each group, t the number of reproduction steps, $v_{j,g}^t$ he velocity vector of agent j in reproduction step of tth, $k_{j,g}^{(t)}$ the position vector of agent j in reproduction step of tth, w the weighting factor, c_1, c_2 the acceleration constants, Rand( ) the random values between 0 and 1, and pbest_j the optimal position vector of agent j.

The variables Rand( ) and Rand( ) are random positive numbers, drawn from a uniform distribution and defined by an upper limit $\text{Rand( )}_{\max}^*$ which is a parameter of the system. In this version, the term variable $w \cdot v_j^{(t)}$ is limited to the range $k_g^{\max}$, for reasons that will be explained below. The values of the elements in gbest_g are determined by comparing the best performances of all the members of i's topological neighborhood, defined by indexes of some other population members, and assigning the best performer's index to the variable g. Thus gbest_g represents the best position found by any member of the neighborhood.

The random weighting of the control parameters in the algorithm results in a kind of explosion as particles velocities and positional coordinates careen toward infinity. Depending on the simulation situationmany coefficients should select to converge for optimal solution.

An important source of the swarm search capability comes from the interactions among particles as they react to one another's findings (analysis calls as inter particle effects).

The value of position vector and velocity vector is determined by the acceleration constants c_1 and c_2. If these values are large, each agent moves on to the target position with high speed and fast variation. If vice versa, agents wander about target place. As weighting factor w is for the searching balance of agent, the value for optimal searching is given by [117]

$$w = w_{\max} - \frac{w_{\max} - w_{\min}}{\text{iter}_{\max}}, \tag{3.6}$$

where $w_{\max}$ is the maximum value of w (0.9), $w_{\min}$ the minimum value of w (0.4), $\text{iter}_{\max}$ the maximum number of iterative, and iter the number of current iterative.

The velocity vector is limited by $v_g^{\min} \le v_{j,g}^{(t)} \le v_g^{\max}$.

3.4 Intelligence by Hybrid System Composed of GA and PSO

3.4.1 Computing Structure of Hybrid System

To compensate the defect in GA and use the advantages of PSO, this study built a hybrid system as shown in Fig. 3.5. In this research, the value of velocity vector for each agent is limited to 1/2 to avoid abrupt variation of position vector. The calculation process for each step is given in Fig. 3.5.

[Step 1] Initialize each variable of GA.

[Step 2] Initialize each variable of PSO.

[Step 3] Calculate affinity of each agent for condition of optimal solution of GA. At this point, the optimal position condition of PSO is introduced into GA.

[Step 4] Arrange the group of PSO and agent in GA as shown in Fig. 3.9.

[Step 5] Update the position vector pbest and the speed vector pbest.

[Step 6] Perform crossover in GA using Euclidian distance and position vector of PSO.

[Step 7] Perform mutation in GA.

[Step 8] If condition of GA is satisfied with the target condition (iteration number or target value), reproduction procedure is halted. Otherwise, it goes to step 3. In Fig. 3.6, IG, ED, PV, and SV refer to initial group, Euclidean distance, position vector, and speed vector, respectively.

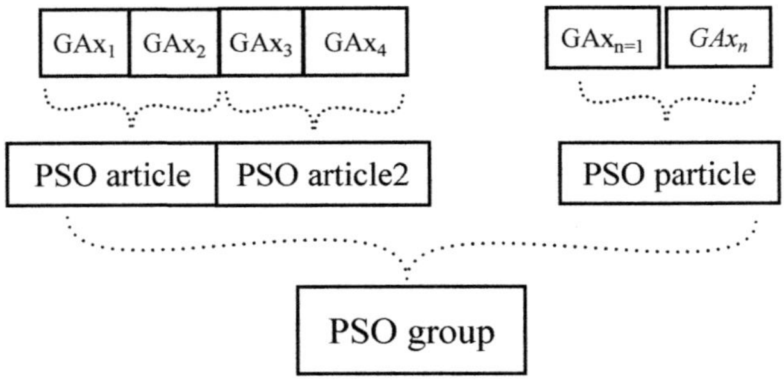

Figure 3.5 Individual structure that combines PSO and GA.

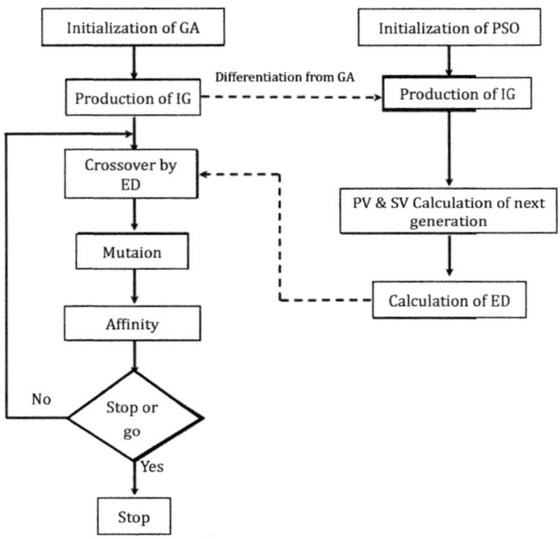

Figure 3.6 Calculation process of the combined GA-PSO algorithm.

In this section, initially, positions of individuals are calculated by Euclidean distance-based method and then mutation and crossover are performed to improve the running speed and to obtain global optimal solutions. Generally, computing procedure in GA is given as [step 1] → [step 6] → [step 7] → [step 8]. However, in this chapter, PSO procedures such as step 3, step 4, and step 5 are added in GA procedure. Therefore, information can be exchanged in computing process. Generally, when we obtain an optimal solution by using GA, because each individual operates separately, they (individuals) cannot give and take information among each other.

Therefore, sometimes optimization can become localization. However, with the hybrid system GA-PSO suggested in this chapter, we can avoid local optimization with each other by genetic algorithm through particle swarm optimization. So, we can obtain total optimization and faster learning time.

3.4.2 The Characteristics of Differentiation Rate of PSO

To prove the learning structure suggested in this section, function $F(x)_1 = \sum_1^3 x_i^2$ is used as a performance function. Figure 3.7 illustrates the contour characteristics of this function. Figure 3.8 illustrates the relationship between the objective function and number of generations, using PSO.

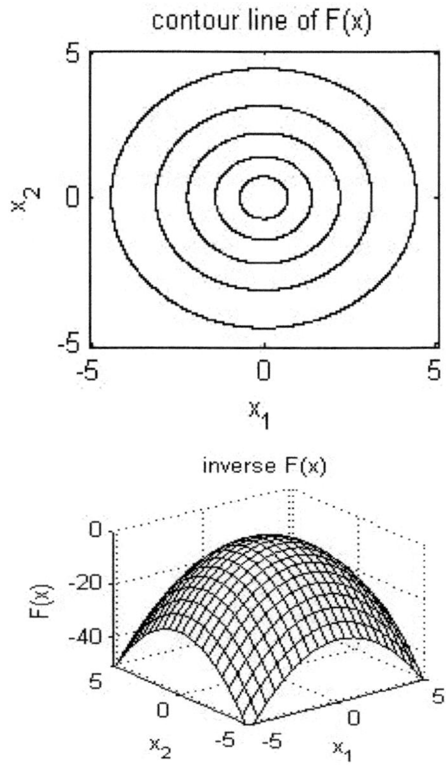

Figure 3.7 Contour of function $F(x) = \sum_{i=1}^3 x_i^2$.

To differentiate the rate, Table 3.3 depicts the relationship between variation of the function and the differentiation rate of PSO. As shown in Fig. 3.8, when the differentiation rate is small, the convergence speed is faster but in the final step, the differentiation rate is large and the convergent speed is faster.

Table 3.3 Performance by differentiation rate

Differentiation rate	x_1 (1.0e-006)	x_2 (1.0e-006)	The value of optimal objective function (1.0e-012)	The value of average objective function (1.0e-008)
PSO - 5	0.3105	−0.4933	0.3398	0.0067
PSO - 10	−0.2799	−0.1014	0.0886	0.1438
PSO - 20	0.1655	0.3842	0.0550	0.0225
PSO - 30	0.0325	0.0197	0.0014	0.0070

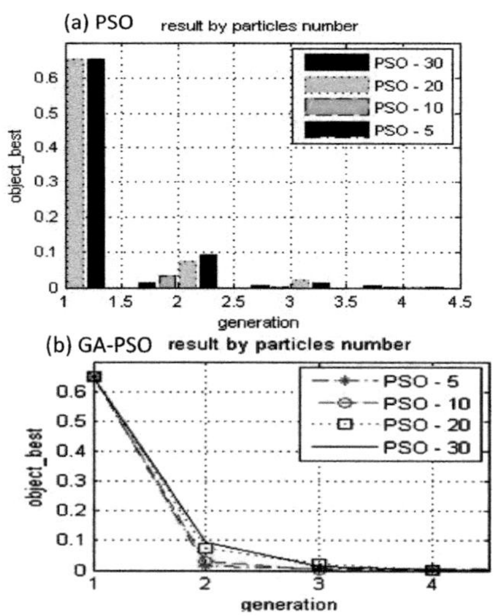

Figure 3.8 The relationship between the objective function and the number of generation in GA, PSO, and GA-PSO.

3.4.3 Characteristics of the Combined GA and PSO

This section shows the characteristics through many ways. Table 3.3 depicts the simulated value of objective function to differentiation rate of PSO.

Figures 3.8 and 3.9 illustrate the relation between GA, PSO, and hybrid GA-PSO. For comparison of both systems, test function, $F(x)_1 = \sum_{i=1}^{3} x_i^2$ and Euclidean data distance are used as shown in Fig. 3.10. This section applies function F_1 to the conventional GA and the hybrid approach (GA-PSO) in order to discuss minimization process.

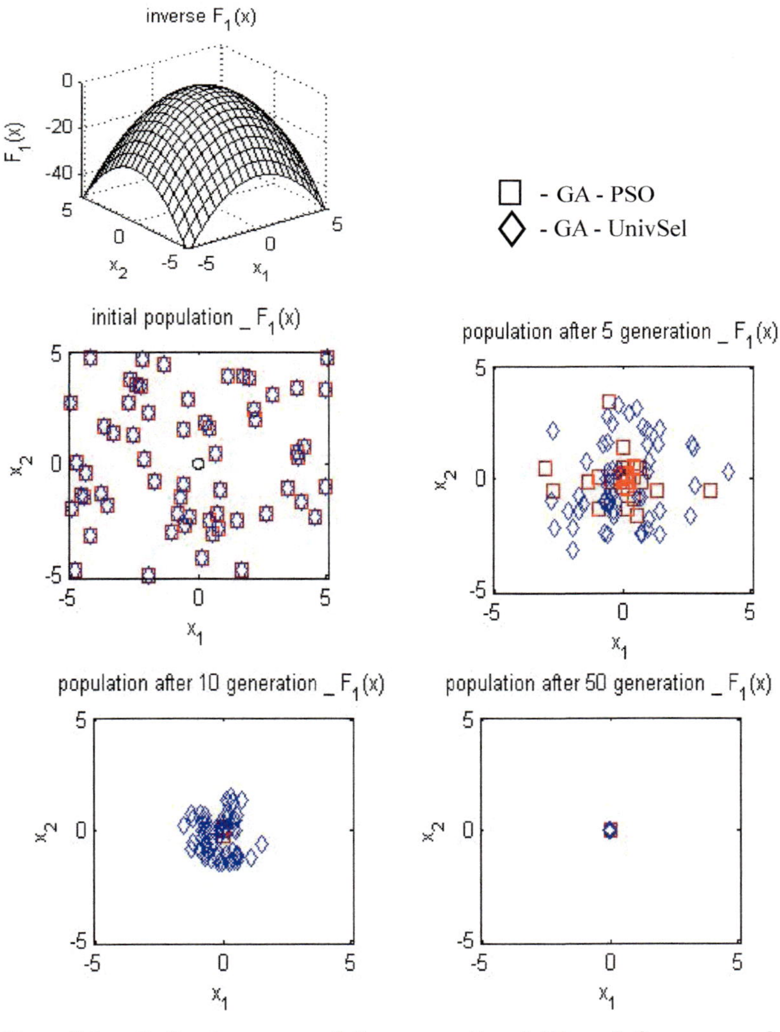

Figure 3.9 Optimal process of the conventional GA and the proposed system GA-PSO by test function $F(x) = \sum_{i}^{3} x_i^2$.

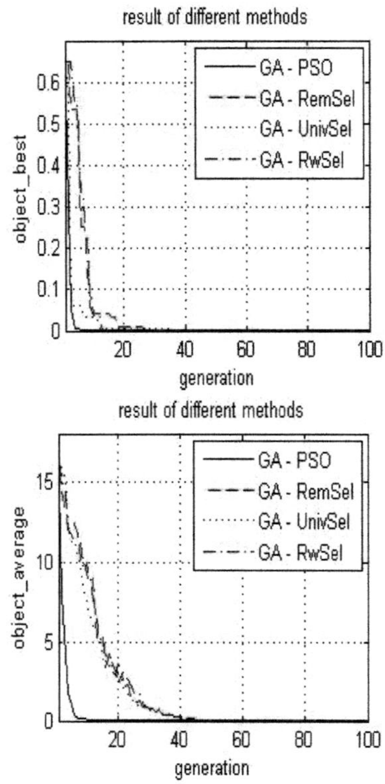

Figure 3.10 Test function $F(x)_1 = \sum_1^3 x_i^2$.

Table 3.4 represents the comparison between the optimal objective function value and the average objective function value of the combined system (GA-PSO) and the GA. The GA-PSO system has the smaller value.

Table 3.4 comparison of combined system (GA + PSO)

	x_1 (1.0e–006)	x_2 (1.0e–006)	The value of optimal objective function	The value of average objective function
GA–PSO	0.0325	0.0197	1.4408e-015	0.0700
GA	–0.2249	0.2585	1.1741e-013	0.1962

3.4.3.1 Rosenbrock function (F_2)

In this section, the Rosenbrock function is used by the hybrid system (GA-PSO). Figure 3.11 illustrates how to obtain optimal solutions for the Rosenbrock function and Table 3.5 depicts the empirical results. Evidently, GA-PSO has better convergence in the search for optimal solutions. Table 3.6 shows the optimal value of the objective function and the average value of the objective function compared in GA and hybrid system GA-PSO with respect to function $F(x)_2$.

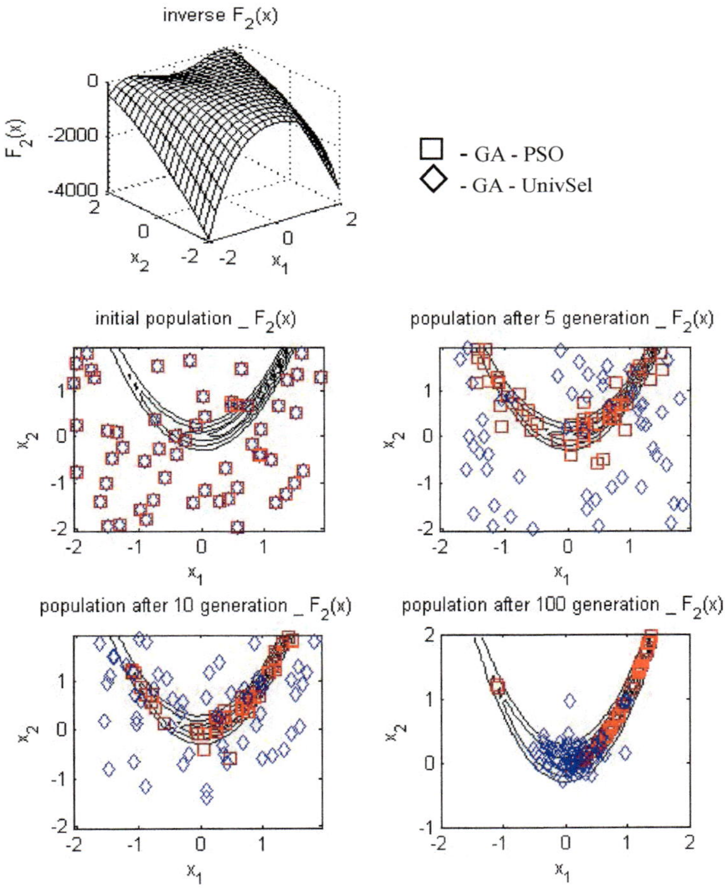

Figure 3.11 (a) Convergence process of the solutions for Rosenbrock function F_2.

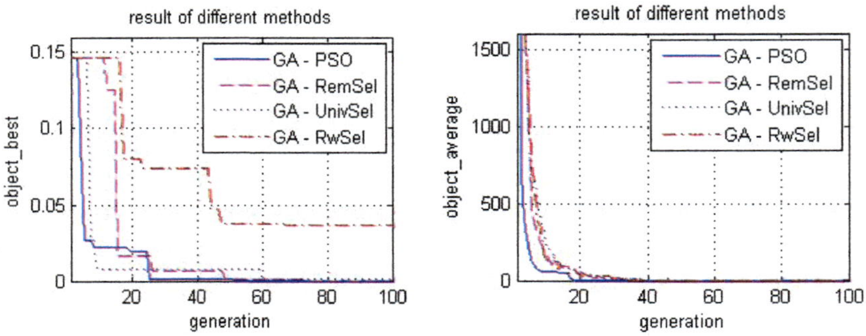

Figure 3.11 (b) Rosenbrock function F_2: convergence characteristics after 100 generations.

Table 3.5 Search space of test functions and initial conditions

Function	Definition		No. of IDs	No. of Re	Rate of DF
	$x_i^{(L)}$	$x_i^{(U)}$			
$F(x)_1 = \sum_{i=1}^{3} x_i^2$	-5.12	5.11	60	100	10
$F_2(x) = 100(x_1^2 - x_2)^2 + (1 - x_1)^2$	-2.048	2.047			
$F_3(x) = (x_1^2 + x_2 - 11)^2 + (x_1 + x_2^2 - 7)^2$	-6	6			
$F_4(x) = \left(0.002 + \sum_{j=1}^{25}\left(j + \sum_{i=1}^{2}(x_i - a_{ij})^6\right)^{-1}\right)^{-1}$	-65.536	65.535			

Note: No. of IDs: number of individuals; No. of Re: number of reproduction; Rate of DF: the rate of differentiation.

3.4.3.2 Himmelblau function (F_3)

Figure 3.12 shows how the proposed method could fasten the convergence for the Himmelblau function. The value of the differentiation rate for comparing the different characteristics in this section is selected to be 10. Figure 3.12a represents the relationship between the objective function and the number of GA generations to the number of particles in PSO. Figure 3.12b compares the existing GA and the proposed hybrid system, EU-GA-PSO. Evidently, the GA-PSO converges much faster than the conventional GA approach. As shown in Fig. 3.12b, in the first step, the conventional GA has faster convergence speed but during the final stages, GA-PSO has more stable speed because GA-PSO searches for optimal solution by incorporating position and direction for search.

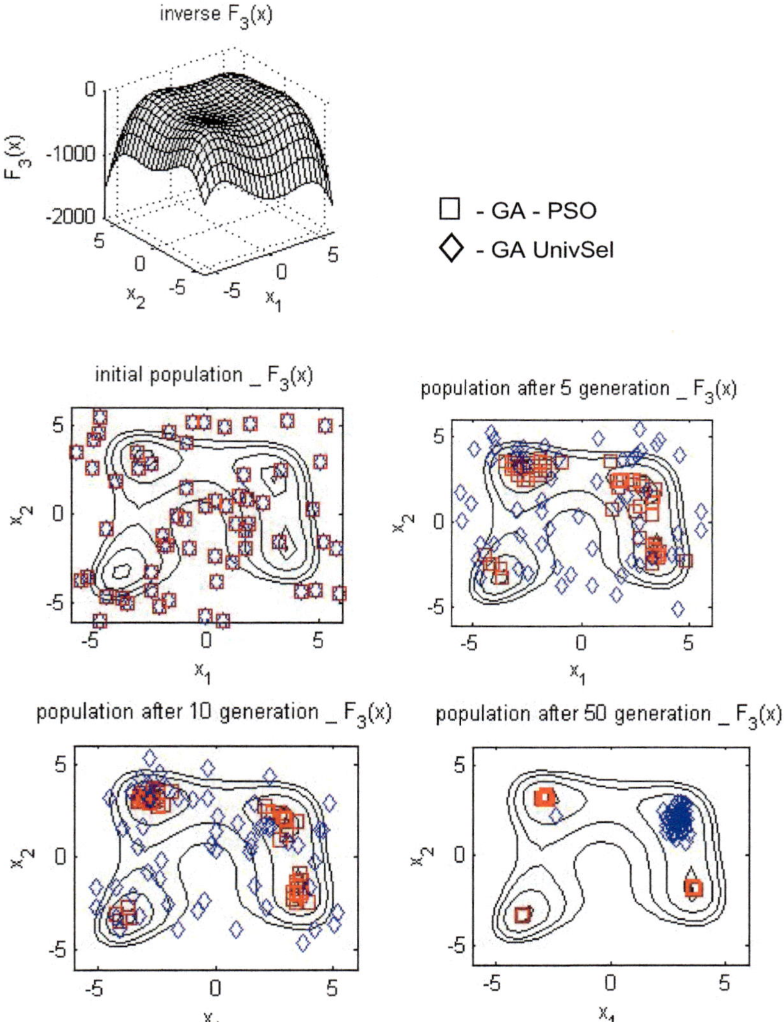

Figure 3.12 (a) Optimal process to Himmelblau function F_3.

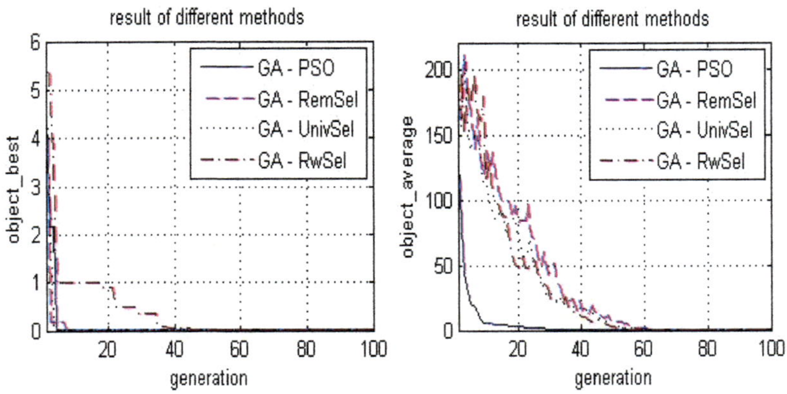

Figure 3.12 (b) Objective function and generation in function Himmelblau F_3.

Table 3.6 Comparison of $F_2(x)$

	x_1	x_2	Optimal value of objective function	Average value of objective function
GA - PSO	1.0026	1.0052	6.7405e-006	2.0807
GA - RemSel	0.9720	0.9447	7.8523e-004	3.0355
GA - UnivSel	0.9612	0.9243	0.0015	5.4145
GA - RwSel	0.8084	0.6540	0.0367	1.2021

Test function $F(x)=\sum_{i=1}^{3} x_i^2$ is defined: $F_3(x)=(x_1^2+x_2-11)^2 + (x_1^2+x_2^2-7)^2$. The GA-PSO depicts better optimal solutions after 5 generations. On the other hand, after 50 generations, GA-PSO represents both optimal solutions (local optimal and global optimal) but it is showing that it is possible to have a local optimal solution because the conventional method has optimal solution at one location. The speed of convergence is illustrated in Fig. 3.12b.

3.4.3.3 Fox hole function *F*(*x*)

Figure 3.13a shows the optimal process to Fox hole F_4 and Fig. 3.13b illustrates the results for the Fox hole function for the different selection schemes namely GA–RemSel, GA–UnivSel, and GA–RwSel.

$$F_4 = \left(0.002 + \sum_{j=1}^{25}\left(j + \sum_{i=1}^{2}\left(x_i - a_{ij}\right)^6\right)^{-1}\right)^{-1} \tag{3.7}$$

3.4.4 Characteristics of Parameter Selection of GA

In GA, in order to transfer gene information of parents or grandfather to offsprings effectively, differentiation is carried out through crossover, reproduction, and mutation. That is, RemSel (Remainder stochastic Sample with replacement Selection), UnivSel (stochastic Universal sampling Selection), and RwSel (Roulette wheel Selection) have been performed. This paper compares and discusses the characteristics of the conventional GA and the proposed hybrid approach (GA-PSO). The results are shown in Fig. 3.13 and the initial condition of the considered test function is also presented in Table 3.7.

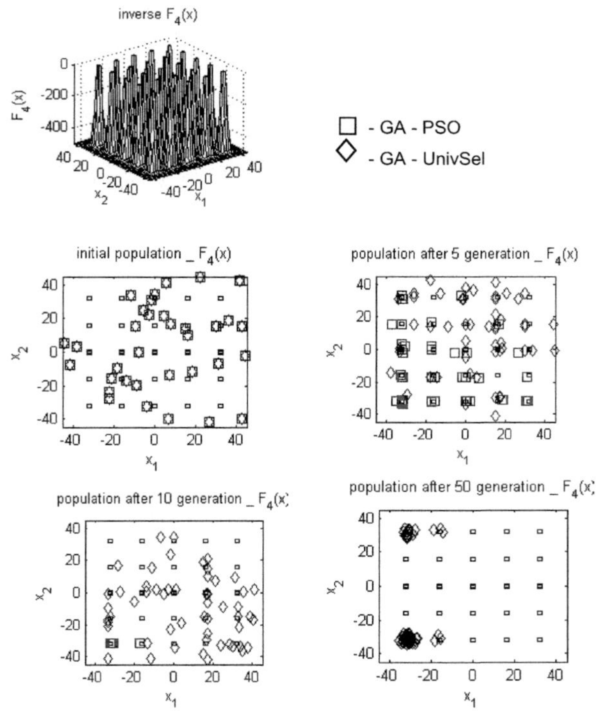

Figure 3.13 (a) Otimal process to Fox hole F_4.

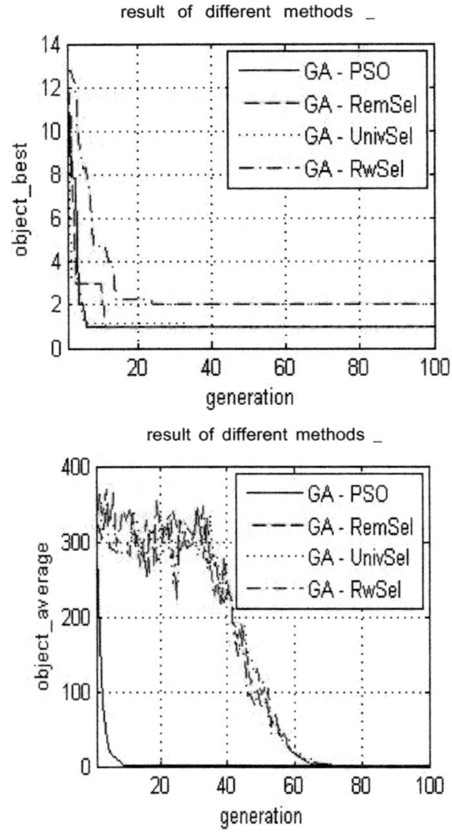

Figure 3.13 (b) Objective function and generation Fox hole F_4.

Table 3.7 Comparison to function F_3

	x_1	x_2	**Optimal value of objective function (1.0e-005)**	**Average value of objective function**
GA - PSO	3.5844	−1.8481	0.0002	0.0285
GA – RemSel	3.0000	2.0000	0.0003	1.1161e-005
GA – UnivSel	2.9998	2.0002	0.1121	2.1361e-005
GA – RwSel	3.0000	2.0000	0.0003	1.0902e-005

3.5 AVR System and PID Controller

This section illustrates application for tuning of PID controller by the hybrid system, GA-PSO. The transfer function of PID controller of the AVR system is given by [100]

$$G(s) = k_p + \frac{k_i}{s} + k_d s, \tag{3.8}$$

and block diagram of the AVR system is shown in Fig. 3.14. The performance index of control response is defined by

$$\begin{aligned}\min F(k_p, k_i, k_d) &= \frac{e^{-\beta \cdot t_s} / \max(t)}{\left(1-e^{-\beta}\right)\left|1-t_r / \max(t)\right|} + e^{-\beta} \cdot M_o + \text{ess} \\ &= \frac{e^{-\beta}\left(t_s + \text{a}_2 \cdot \left\|1-t_r/\max(t)\right| \cdot M_o\right|\right)}{\left(1-e^{-\beta}\right)\left|1-t_r/\max(t)\right|} + \text{ess} \\ &= \frac{e^{-\beta}\left(t_s/\max(t) + \alpha \cdot \text{M}_o\right)}{\alpha} + \text{ess}\end{aligned} \tag{3.9}$$

$\alpha = (1-e^{-\beta}) \cdot \left|1-t_r/\max(t)\right|$, k_p, k_i, k_d are the parameters of the PID controller, β the weighting factor, M_o overshoot, t_s the settling time (2%), ess the steady-state error, and t the desired settling time.

Table 3.8 The results of function F_4

	x_1	x_2	Optimal value of objective function (1.0e-005)	Average value of objective function
GA - PSO	-31.9783	-31.9783	0.9980	0.9980
GA - RemSel	-31.9764	-31.9833	0.9980	0.9980
GA - UnivSel	-31.9788	-31.9790	0.9980	0.9980
GA - RwSel	-15.9861	-31.9702	1.9920	1.9920

In Eq. (3.9), if the weighting factor, β, increases, rising time of response curve is small, and when β decreases, rising time is large.

The performance criterion is defined as M_o = 50.61%, ess = 0.0909, t_r = 0.2693 (s), and t_s = 6.9834 (s).

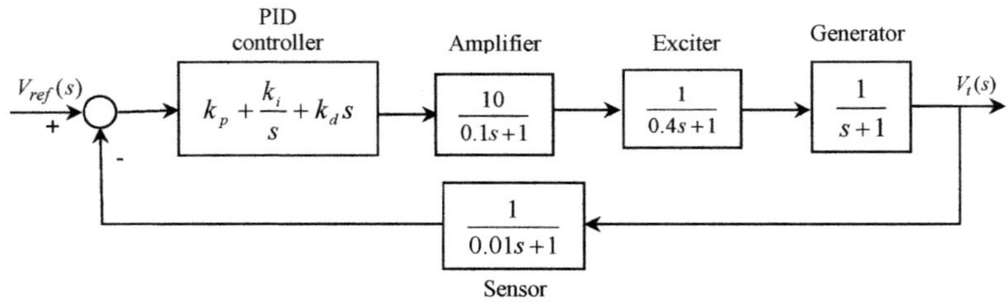

Figure 3.14 Block diagram of an AVR system with a PID controller.

3.5.1 Operation of PSO in the EU-GA-PSO System

In PSO, parameters k_p, k_i, and k_d are defined and the number of each age is fixed as 10 with the number of groups 5.

Here, weighting factor: $w_{\max} = 0.9$, $w_{\min} = 0.4$, restriction of velocity vector: $V_{k_p}^{\max} = k_p^{\max}/2$, $V_{k_i}^{\max} = k_i^{\max}/2$, $V_{k_d}^{\max} = k_d^{\max}/2$, $V_{k_p,k_i,k_d}^{\min} = -V_{k_p,k_i,k_d}^{\max}$, acceleration constant: $c_1 = 2$, $c_2 = 2$.

3.5.2 EU-GA-PSO

EU-GA-PSO mentioned above is applied to PID controller tuning. The detailed description is as follows.

3.5.3 Characteristics of Variation of β in EU-GA-PSO

Terminal voltage step response of an AVR system with controller is given as Figs. 3.15 and 3.16. The characteristics of the AVR system obtained due to the variation of β using the EU-GA-PSO approach is shown in Fig. 3.15.

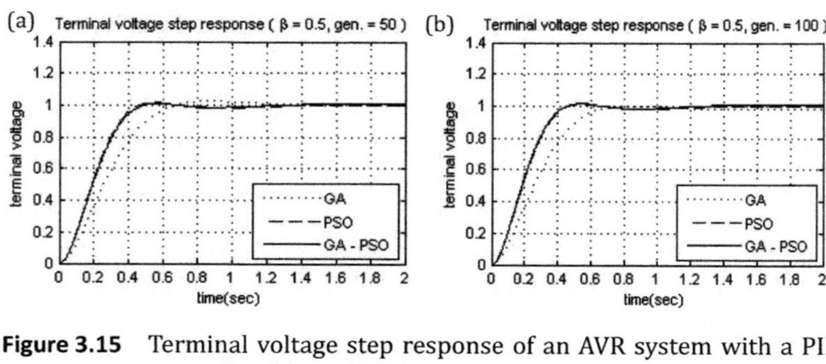

Figure 3.15 Terminal voltage step response of an AVR system with a PID controller (β = 0.5, (a): generations = 50, (b): generations = 100).

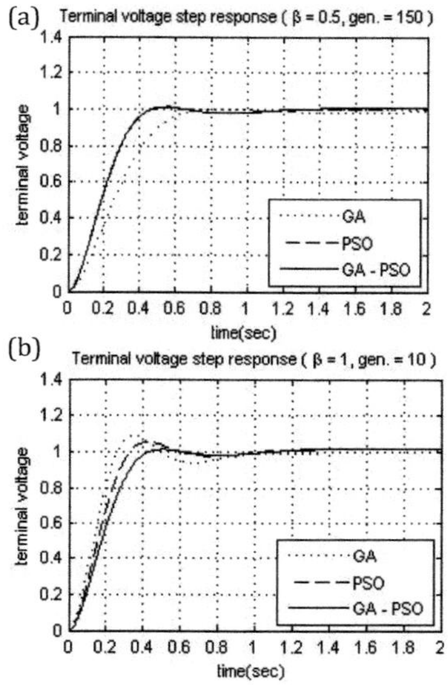

Figure 3.16 Terminal voltage step response of an AVR system with PID controller ((a): β = 0.5, generations = 150, (b): (β = 1, generations = 10)).

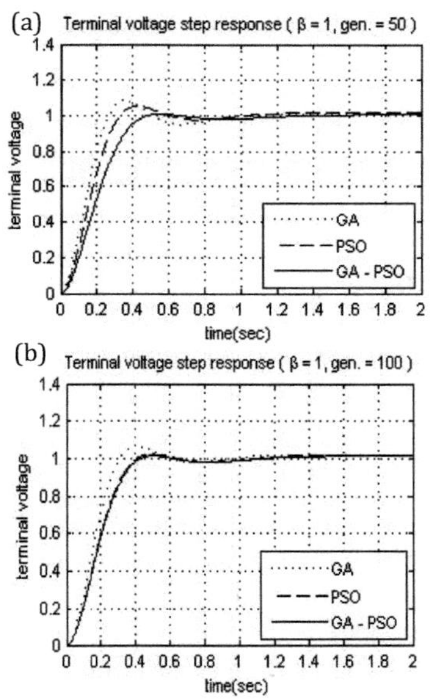

Figure 3.17 Terminal voltage step response of an AVR system with a PID controller ((a) β = 1, generations = 50, (b) (β = 1, generations = 100).

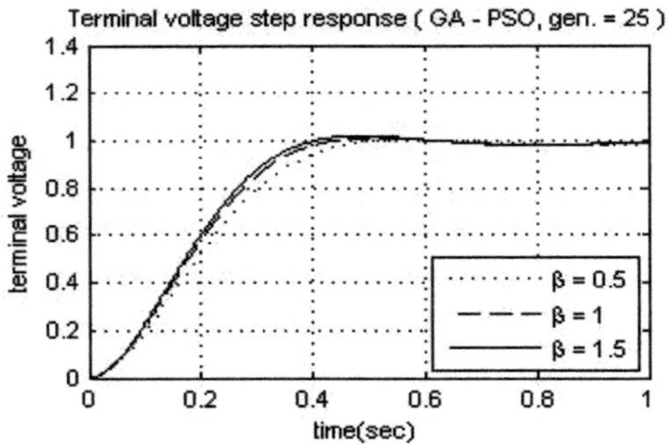

Figure 3.18 Terminal voltage step response of an AVR system with the GA-PSO PID controller.

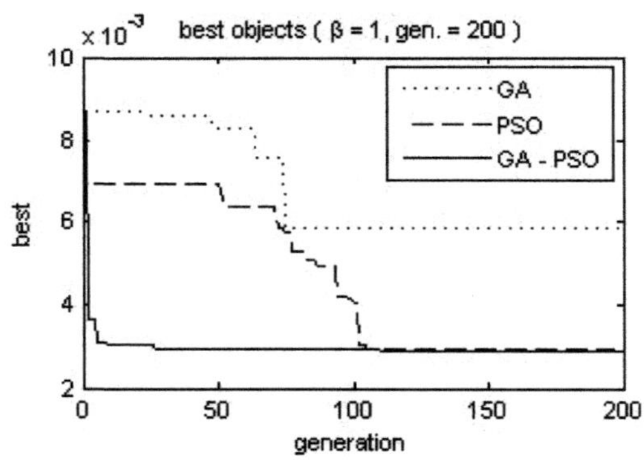

Figure 3.19 Comparison of the best objects values using all the methods (β = 1.5, generations = 200).

3.6 Summary

Even though several deep control theories have been developed, the proportional-integral-derivative (PID) controllers have been widely used owing to their simple structure, which can be easily understood and implemented for a wide range of process control, motor drives, flight control, and instrumentation. Owing to their popularity in the industrial world, over the past 50 years, several approaches for determining PID controller parameters have been developed for stable processes that are suitable for auto-tuning and adaptive control and for single input single output (SISO) systems. In spite of the enormous amount of research work reported in the tuning approaches, many PID controllers are poorly tuned in practice. One of the reasons is that most of the tuning methods are derived for particular processes and situations and therefore apply well only to their own areas. It is a common experience that we are not certain which tuning method should be chosen to provide good control for a given process. Intelligent controllers could self-initialize and recalibrate even with little a priori knowledge especially due to the occurrence of significant changes in the process dynamics. In this chapter, GA Euclidean–based distance concept and PSO were introduced for the enhancement of optimal tuning of conventional PID controllers. By incorporating the Euclidean distance measure for selecting mutation or crossover points, the search space is well explored. Therefore, GA can provide exact optimal solution, while it can avoid local optimal solutions. On the other hand, this chapter also proposed a hybrid PSO approach to enhance the speed of convergence. A candidate solution for a specific problem in GA is called an individual or a chromosome and consists of a linear list of genes. Each individual represents a point in the search space, and hence a possible solution to the problem. A population consists of a finite number of individuals. Each individual is decided by an evaluating mechanism to obtain its fitness value. Based on this fitness value and undergoing GA-PSO operators with Euclidean distance, a new population is generated iteratively with each successive population referred to as a generation. Experiment results reveal the efficiency of the proposed approach with a faster convergence and optimal solutions. The GA-PSO system proposed in this chapter for PID controller tuning could be easily extended to model other complex problems involving local optimal and global optimal solutions.

Table 3.8 is simulation results of GA, PSO, and GA-PSO at generation variation from generation 50 to 200 and performance index to β = 0.5 and 1 to obtain optimal value of PID parameters $K_pK_iK_d$. Table 3.9 shows the tuned parameters of PID controller by beta value and generation.

Table 3.9 Simulation results of PID controller in AVR system to β variation

β	No. of generations	k_p	k_i	k_d	M_o (%)	ess	t_s	t_r	Objective
0.5	25	0.6204	0.4929	0.2232	0.97	0.0097	0.4570	0.2973	0.0079
1	25	0.6584	0.5819	0.2548	1.71	0.0166	0.4000	0.2651	0.0030
1.5	25	0.6801	0.6260	0.2681	1.97	0.0186	0.3770	0.2523	0.0072

Table 3.10 Comparison of the evaluation value between both methods (β = 0.5~1.0, generation = 50~200)

	Number of generation	Type of controller	k_p	k_i	k_d	Mo(%)	ess	t_s	t_r	Evaluation value
β = 0.5	50	GA	0.5045	0.3654	0.1400	2.3004	0.0005	1.0010	0.4136	0.0176
		PSO	0.6572	0.4816	0.2284	1.4589	0.0035	0.4280	0.2842	0.0077
		GA-PSO	**0.6310**	**0.4929**	**0.2232**	**0.8857**	**0.0083**	**0.4480**	**0.2941**	**0.0077**
	100	GA	0.5045	0.3080	0.1400	0.2261	0.0148	0.6590	0.4231	0.0108
		PSO	0.6572	0.4816	0.2284	1.4589	0.0035	0.4280	0.2842	0.0077
		GA-PSO	**0.6554**	**0.5224**	**0.2374**	**1.3606**	**0.0095**	**0.4190**	**0.2779**	**0.0075**
	150	GA	0.5045	0.3142	0.1416	0.2254	0.0128	0.6570	0.4214	0.0108
		PSO	0.6537	0.4852	0.2292	1.3115	0.0045	0.4300	0.2845	0.0076
		GA-PSO	**0.6447**	**0.5060**	**0.2350**	**0.8581**	**0.0086**	**0.4300**	**0.2825**	**0.0073**
	200	GA	0.5061	0.3080	0.1420	0.0044	0.0149	0.6600	0.4211	0.0106
		PSO	0.6491	0.4915	0.2317	1.0493	0.0059	0.4300	0.2839	0.0075
		GA-PSO	**0.6447**	**0.5058**	**0.2350**	**0.8564**	**0.0085**	**0.4300**	**0.2825**	**0.0073**
β = 1	50	GA	0.9186	0.8100	0.3935	8.7179	0.0122	0.8550	0.1758	0.0083
		PSO	0.7893	0.7196	0.3105	5.2491	0.0154	0.8320	0.2155	0.0069
		GA-PSO	**0.6415**	**0.4825**	**0.2274**	**0.8820**	**0.0055**	**0.4400**	**0.2889**	**0.0030**

Number of generation	Type of controller	k_p	k_i	k_d	Mo(%)	ess	t_s	t_r	Evaluation value
100	GA	0.8326	0.8100	0.3277	6.8331	0.0171	0.5570	0.2037	0.0058
	PSO	0.6834	0.6096	0.2611	2.2200	0.0164	0.5340	0.2559	0.0040
	GA-PSO	**0.6657**	**0.5697**	**0.2548**	**1.4503**	**0.0143**	**0.3980**	**0.2639**	**0.0029**
150	GA	0.8326	0.8100	0.3277	6.8331	0.0171	0.5570	0.2037	0.0058
	PSO	0.6651	0.5690	0.2533	1.4511	0.0142	0.3990	0.2649	0.0029
	GA-PSO	**0.6523**	**0.5189**	**0.2398**	**1.0510**	**0.0094**	**0.4200**	**0.2773**	**0.0029**
200	GA	0.8326	0.8100	0.3277	6.8329	0.0171	0.5570	0.2037	0.0058
	PSO	0.6660	0.5682	0.2543	1.4285	0.0140	0.3980	0.2641	0.0029
	GA-PSO	**0.6522**	**0.5188**	**0.2398**	**1.0472**	**0.0094**	**0.8680**	**0.2773**	**0.0029**

Chapter 4

Intelligent Vector Control Using a Hybrid GA-PSO System

4.1 Overview

Induction motors have been widely used in various industries as actuators or drivers to produce mechanical motions and forces. Since it is estimated that more than around 50% of the world's electric energy generated is consumed by electric machines, improving efficiency in electric drives is important [132, 133]. That is mainly, for two reasons: economy and environmental pollution reduction [135, 136]. Generally, induction motors require both wide operating range of speed and fast torque response in operational conditions, regardless of load variations. They have high efficiency at rated speed and torque. Therefore, they have been widely used as drivers in industry. However, at light loads, iron losses increase dramatically, which need to be reduced considerably for efficiency [137, 138]. To improve the motor efficiency, the flux must be reduced, obtaining a balance between copper and iron losses [139]. There are two different approaches to improving the efficiency in an induction motor: the power measure–based method and the loss model–based method.

In the first approach, an optimum flux search algorithm is used, and the drive power consumption is measured [140–142]. This approach does not require the information or knowledge

Tuning Innovation with Biotechnology
Dong Hwa Kim

ISBN 978-981-4745-35-2 (Hardcover), 978-1-315-36458-2 (eBook)
www.panstanford.com

of motor parameters. However, it is only efficient in the steady-state condition and for transients, and a controller actuating as a supervisor is needed to optimize efficiency with reduced flux levels. Some results propose different optimum flux search algorithms in an indirect field-oriented control and compared in a standard induction motor [133].

In the second approach, when a motor loss model is available, the loss minimization optimum flux can be computed analytically [134, 135]. The main advantage of this method is the simplicity of the approach — it does not require extra hardware. However, it is necessary to obtain an accurate knowledge of motor parameters, which change considerably with temperature, saturation, skin effect, variation of load, etc. In this method, the optimum tools such as dynamic programming let us solve the problem of efficiency optimization to calculate the optimum flux trajectories. However, this approach presents the difficulty to be implemented and it is also very sensitive to parameter changes.

To meet this requirement, many control approaches such as intelligent optimization are introduced into loss minimization control or energy saving control area in an induction motor. The field-oriented control is one of the most successful in meeting the above requirements. Due to advances in power electronics and DSP technology, variable-speed drives of induction motors using the field-oriented control have been widely used in many applications, such as ac servo, electric vehicle drive systems, and so on [2–5]. Using the field-oriented control, a highly coupled, nonlinear, multivariable induction motor can be simply controlled by means of implementing linear independent decoupled control of torque and flux, similar to separately excited dc motors. Therefore, high-performance torque control requires fast enough current response for the current regulator to track the reference current. However, due to limitations of voltage and current ratings on the inverter DC link, the input voltage and current of an induction motor are limited accordingly. Hence, the developed torque in the motor should be limited for low loss operation under these input constraints.

In particular, since the objective of a variable-speed control system for higher efficiency is to track the reference speed as fast as possible, under the constraints of input voltage and current, control approaches which yield the maximum torque over the entire speed range can be usefully applicable to minimum-time speed control of induction motors. However, most researchers who deal with the speed control of induction motors have not discussed the maximum-torque generation scheme, under low loss consideration.

Results suggest a voltage control strategy for the maximum-torque operation of induction motors in the field-weakening region, considering the voltage and current constraints, not mentioning the stator resistor. In order to produce the maximum torque, flux current components are compensated, as much as current components omitted by neglecting the stator resistor. Hence, the resultant maximum-torque generation algorithm is not the theoretically exact solution to maximize torque, because the solution is obtained by using an approximated formulation. Some results suggest the entire current input is used for generating the rotor flux before the torque is developed. However, at the moment that the rotor flux reaches the steady-state value, since the entire input current is switched to produce torque current component, this research can guarantee the instantaneous maximum torque, but a constant time interval is required for producing the rotor flux initially. Hence, the minimum-time speed control is not guaranteed and also cannot provide low loss control strategy.

In this chapter, a new minimum-time minimum-loss control algorithm for induction motors using hybrid system GA-PSO is suggested to obtain high performance, as well as high efficiency, under practical constraints on voltage and current. The validity of the suggested scheme, which carries out minimum-time speed control in the transient state and minimum-loss control in the steady state, will be revealed via simulation, including an induction motor model.

In the remainder of this chapter, the dynamical model for vector control is overviewed. This chapter gives a brief overview for the vector control of an induction motor and describes the structure of indirect vector control, and presents PI speed controller and PI current controller tuning algorithm using GA-PSO. Of course, description will give the simulation and experiment of the suggested speed and current control scheme. Finally, conclusions and further studies are explained in the last section.

4.2 Mathematical Model of Induction Motor for GS-PSO

The control algorithm implemented in a study for vector is a rotor flux-orientated control strategy, based on the field-orientated control (FOC) structure [136]. This strategic operation controls the instantaneous direct torque and flux controls by means of coordinate transformations by the PI regulators when the position of the rotor flux and two-phase currents is given. Therefore, a really accurate and efficient motor control is achieved. The generic FOC structure needs to be augmented with two modules in order to address the asynchronous drive specificity. The current model must be added to the generic structure. It takes as input both i_d and i_q current as well as the rotor mechanical speed and gives the rotor flux position as the output [132, 133, 135]. These two components of the phase currents feed the transformation module and provide gives the current in the d, q rotating reference frame, whose i_d and i_q components are compared to the flux reference, i_{sdref} and the torque reference, i_{sdref}. The torque command i_{sdref} corresponds to the output of the speed regulator, and the flux command i_{sdref} is the right rotor flux command for every speed reference. The current regulator outputs are v_{sdref} and v_{sdref}. The outputs of this projection are v_{sdref} and v_{sdref}, the components of the stator vector voltage in the α, β orthogonal reference frame. The speed controller of the AC induction machine has two roles: low-speed range, in which the motor speed is below the nominal speed, and the high-speed range, in which the motor speed is higher than the nominal speed. The back electromotive force depends on both the motor speed and the rotor flux. Therefore, there are two limitations—torque production and drive efficiency—due to problems with magnetic saturation and heat dissipation. The rotor flux should be maintained at its nominal value during the low-speed operation so as to achieve the highest mutual torque production. However, it must be reduced in the high-speed operation in order to avoid magnetic saturation and the generation of too high back electromotive force. Reducing the rotor flux in such a way can extend the high efficiency operating range of the drive. This functionality is integrated into the field-weakening module. Where vector control is importance and for these flux controls, angle ρ between the rotor vector and the field-oriented vector is defined as

$$\rho = \tan^{-1}\frac{\psi_b}{\psi_a}, \tag{4.1}$$

where, ψ_a, ψ_b are flux vectors in state space.

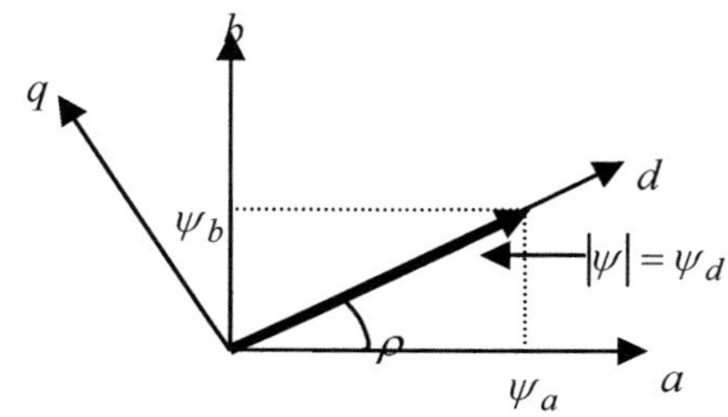

Figure 4.1 Stator frame and field-oriented frame.

This approach needs the rotor resistance as a parameter. Accurate knowledge and representation of the rotor resistance is essential to achieve the highest possible efficiency from the control structure. The transformation for vector control mentioned above is generally given as [134, 137, 138]

$$\begin{bmatrix} i_d \\ i_q \end{bmatrix} = \begin{bmatrix} \cos\rho & \sin\rho \\ -\sin\rho & \cos\rho \end{bmatrix} \begin{bmatrix} i_a \\ i_b \end{bmatrix}, \tag{4.2}$$

$$\begin{bmatrix} \psi_d \\ \psi_q \end{bmatrix} = \begin{bmatrix} \cos\rho & \sin\rho \\ -\sin\rho & \cos\rho \end{bmatrix} \begin{bmatrix} \psi_a \\ \psi_b \end{bmatrix}. \tag{4.3}$$

Since rotor flux is $|\psi| = \psi_d = \sqrt{\psi_a^2 + \psi_b^2}$ and $\cos\rho = \frac{\psi_a}{|\psi|}$, $\sin\,\rho = \frac{\psi_a}{|\psi|}$, the following equation is obtained:

$$\begin{aligned} i_d &= \frac{\psi_a i_a + \psi_b i_b}{|\psi|} \\ i_q &= \frac{\psi_a i_b - \psi_b i_a}{|\psi|}. \\ \psi_d &= \sqrt{\psi_a^2 + \psi_b^2} = |\psi| \\ \psi_q &= 0 \end{aligned} \tag{4.4}$$

The state equation for field-oriented control is

$$x_e = (\omega, i_q, \psi_d, i_d, \rho)^T. \tag{4.5}$$

The state transformed in stator vector for field-oriented control is defined by

$$\begin{aligned}
i_q &= \frac{\psi_a i_b - \psi_b i_a}{\psi_d} \\
\psi_d &= \sqrt{\psi_a^2 + \psi_b^2} \\
i_d &= \frac{\psi_a i_a + \psi_b i_b}{\psi_d} \\
\rho &= \tan^{-1}\left(\frac{\psi_b}{\psi_a}\right).
\end{aligned} \tag{4.6}$$

The final equation for field-oriented vector control is defined by

$$\begin{aligned}
\frac{d\omega}{dt} &= \mu\psi_d i_q - \frac{T_L}{J} - \frac{B}{J}\omega \\
\frac{di_q}{dt} &= -\gamma i_q - n_p\beta\omega\psi_d - n_p\omega i_d - \alpha M\frac{i_q i_d}{\psi_d} + \frac{1}{\sigma}u_q \\
\frac{d\psi_d}{dt} &= -\alpha\psi_d + \alpha M i_d \\
\frac{di_d}{dt} &= -\gamma i_d + \alpha\beta\psi_d + \alpha M\frac{i_q^2}{\psi_d} + \frac{1}{\sigma}u_d, \\
\frac{d\rho}{dt} &= n_p\omega + \alpha M\frac{i_q}{\psi_d}
\end{aligned} \tag{4.7}$$

where

$$\sigma = L_s\left(1 - \frac{M^2}{L_s L_r}\right), \ \mu = \frac{n_p M}{J L_r}, \ \alpha = \frac{R_r}{L_r}, \ \beta = \frac{M}{\sigma L_r},$$

$$\gamma = \frac{R_r M^2}{\sigma L_s L_r^2} + \frac{R_s}{\sigma L_s} \tag{4.8}$$

μ is the torque constant and α the inverse of the rotor constant L_r/R_r.

In field-oriented equation, the input to stator is u_q and stator voltage u_d, q coordinate i_q of stator current to generate speed, w and d coordinate of i_d as stator voltage u_d to generate rotor flux ψ_d [132, 133].

4.3 Vector Control Overview

The direct torque and flux control for induction machine drives has been developed as direct torque control (DTC) in [132, 134] and as a direct self control (DSC) in [133]. The technique was generalized to all ac drives, as torque vector control (TVC) in [133], and it was recognized as a viable alternative to FOC [4]. Industrial drives with DTC are present on the market today [135]. DTC abandons the stator current control philosophy, characteristic of FOC, and achieves bang-bang torque and flux control by directly modifying the stator voltage in accordance with the torque and flux errors. DTC is characterized by fast dynamic response, structural simplicity, and strong robustness in the face of parameter uncertainties and perturbations. It does not employ current controllers and pulse width modulation (PWM), and it is well suited for sensorless drives. The variable structure control approach [141–143] is effective for torque and flux control as high-frequency switching control strategy for nonlinear systems with uncertainties. It features good robustness in the face of parameter uncertainties and other disturbances because it has fast response control [143]. The direct torque and flux control can be implemented as VSC (variable structure control). The fast dynamic response of classic DTC is entirely preserved, while the steady-state behavior is significantly improved. Comparative experimental results illustrate the characteristic behavior of each scheme and confirm their feasibility. Electrical machines normally operate at rated flux in a variable frequency drive to get best transient response. However, most of the time, drives operate with light loads. In this case, the core losses become excessive causing poor efficiency [134]. If significant energy savings are to be obtained, it is necessary to optimize the efficiency of motor drive systems. Various methods have been investigated to achieve this purpose [2–4]. In [132], an optimal controller scheme has been proposed and implemented on a conventional FOC structure of an induction motor. Simulation results have shown that it is possible to significantly improve the induction motor drive efficiency when it is slightly loaded. Many existing schemes or modern control techniques use to achieve high-performance torque and flux control. All approaches that realize the motor control by direct modification of stator voltage can be based on torque and flux errors as DTC schemes [135–201]. The DTC has many advantages—one of them is that simultaneous and decoupled control of torque and stator flux is achieved by direct adjustment of the stator voltage, in accordance with the torque and flux errors, without intermediate current control or decoupling network. Generally, in sensorless DTC scheme, it is composed of four components such as the torque controller, the stator flux controller, the switching strategy, and a state observer. Classic DTC uses hysteresis torque, flux controllers, and a switching table that generates the VSI switching signals [135]. On the basis of the torque and flux errors, the VSI voltage vector is determined so as to rapidly reduce these errors and to maintain them within prefixed limits. The indirect field-orientated-control (IFOC) approach is one of the powerful ways to guarantee the dynamic performance of induction motor drives. It has been estimated in IFOC based on a real time model of the induction motor. As the stator windings' influence cannot be neglected anymore due to the low switching frequencies in high-power systems, a parameter adaptation in the vector controller using the modified reactive power is needed through simulation and experiment.

This introduces a cross coupling effect we can eliminate by introducing decoupling terms in each axis of the synchronously rotating reference frame. The IFOC method associated with compensation scheme of parameter variation is efficient enough to achieve good dynamic performances of an induction machine drive, due to its simple implementation and its physical approach of the process. Recently practical investigations are performed on an experimental drive system to validate the IFOC method in a high-power application.

4.4 Structure of Indirect Vector Control

As the vector controlled induction machine is assumed to be current fed from an ideal current controlled PWM inverter, operation with constant, rated flux command would be discussed. As the indirect vector controller is the scheme composed of the appropriate decoupling circuit for each of the three orientation possibilities such as stator, air-gap, and rotor flux–oriented control, it incorporates only PI speed controller. Decoupling circuits neglect iron loss, magnetic saturation, and resistance variations and have the well-known form, representation of the induction machine, in terms of space vectors. That is, the indirect vector control system neglects the core loss. The electrical torque in an induction machine can be expressed as

$$T_e = \frac{3}{2}\frac{P}{2}\frac{L_m}{L_r}\lambda^e_{\mathrm{dr}}i^e_{\mathrm{qs}}, \tag{4.9}$$

where rotor flux instantaneous speed ω_{e}, $\lambda^e_{\mathrm{r}} = \lambda^e_{\mathrm{dr}} + j\lambda^e_{\mathrm{qr}} = \lambda^e_{\mathrm{dr}}$, $\lambda^e_{\mathrm{qr}} = 0$, $i^e_{\mathrm{qs}} = -\frac{L_r}{L_m} i^e_{\mathrm{qr}}$.

The slip equations for an induction motor in an arbitrary synchronously rotating reference frame are given by

$$\begin{aligned}\omega_e - \omega_r = \omega_{\mathrm{sl}} &= -\frac{R_r i^e_{\mathrm{qr}}}{\lambda^e_{\mathrm{dr}}} = \frac{R_r}{\lambda^e_{\mathrm{dr}}}\frac{L_m}{L_r}i^e_{\mathrm{qs}} \\ &= \left[\left(1 + p\frac{L_r}{R_r}\right)\frac{1}{i^e_{\mathrm{ds}}}\right]\frac{R_r}{L_r}i^e_{\mathrm{qs}}.\end{aligned} \tag{4.10}$$

where, i^e_{qs} is torque current, i^e_{ds} is flux current. When i^e_{ds} and i^e_{qs} is decided by ω_{sl}, rotor flux position θ_{e} is given by

$$\theta_e = \int_0^t \omega_e d\tau = \int_0^t (\omega_r + w_{\mathrm{sl}})d\tau \tag{4.11}$$

current and slip angle ω_{sl} through θ_{e} is controlled, then $\lambda^e_{\mathrm{qr}} = 0$ [132].

4.5 GA-PSO for Vector Control

This section focuses on the advantage of PSO into the mutation process of GA, for improving the GA learning behavior. Euclidean distance is used on crossover to avoid local optimal and obtain fast running time of solution. We illustrate the performance of the method using four test functions described in Chapter 3.

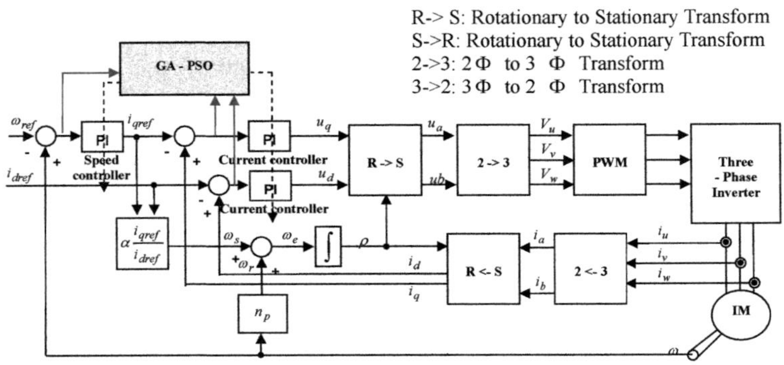

Figure 4.2 Block diagram of indirect vector PI controller using GA-PSO.

4.6 Simulation and Discussion

The figures in Chapter 3 represent convergence contour and convergence characteristics after 100 generations of the GA-PSO approach incorporating the Euclidean distance obtained on Rosenbrock function. Experiment results reveal the efficiency of the proposed approach with a faster convergence and optimal solutions.

The GA-PSO system proposed in this chapter for PID controller tuning could be easily extended to model other complex problems involving local optimal and global optimal solutions.

For more effective simulation, Fig. 4.2 shows the block diagram of indirect vector PI controller using GA-PSO. This chapter uses three-phase squirrel induction motor model with rotary encoder, 2.2 KW, 3 HP, 220/380 V, 60 Hz, 8.5/5 A provided by LG-OTIS and its parameters are shown in Table 4.1. To observe the characteristic of motor and performance of GA-PSO-based controller on the motor mentioned above, we simulate with MATLAB.

The initial value of GA-PSO to obtain optimal PI parameters is shown in Table 4.2 and parameters of GA-PSO are defined as follows:

Members of each agent: Kp_speed, Ki_speed, Kp_current, Ki_current.

- The number of group: 3

$$V^{\max}_{\text{Kp(Ki)\_speed,Kp(Ki)\_current}} = \frac{1}{2}\text{Kp(Ki)}^{\max}_{\text{\_speed,current}},$$

$$V^{\min}_{\text{Kp(Ki)\_speed,Kp(Ki)\_current}} = -V^{\max}_{\text{Kp(Ki)\_speed,Kp(Ki)\_current}}.$$

- Acceleration constant: $c_1 = c_2 = 2$.
- Max generation: 20.
- Weighting factor: $W_{\max} = 0.9$, $W_{\min} = 0.4$.
- Limit of speed vector:

Table 4.1 Parameter initial values of the induction motor

Parameter	Value
Number of pole pair (n_p)	2
Stator resistance (R_s)	0.84 [Ω]
Rotor resistance (R_r)	0.3858 [Ω]
Stator inductance (s)	70.6 [mH]
Rotor inductance R_r	70.6 [mH]
Mutual inductance (M)	67.2 [mH]
Rotor mass moment of inertia (J)	0.008 [kg·m^2]
Coefficient of friction (B)	0.004 [N·m·s/rad]

Table 4.2 Initial values of induction motor parameters for simulation

PI parameters	Range	
	Min	Max
Kp_speed	0	1
Ki_speed	0	1
Kp_current	0	20
Ki_current	0	200

Figure 4.3 shows speed tracking curve, flux amplitude tracking, d-axis current, and q-axis current by indirect vector PI controller based on GA-PSO. In the figure, detailed difference cannot be seen but Table 4.3 shows the difference.

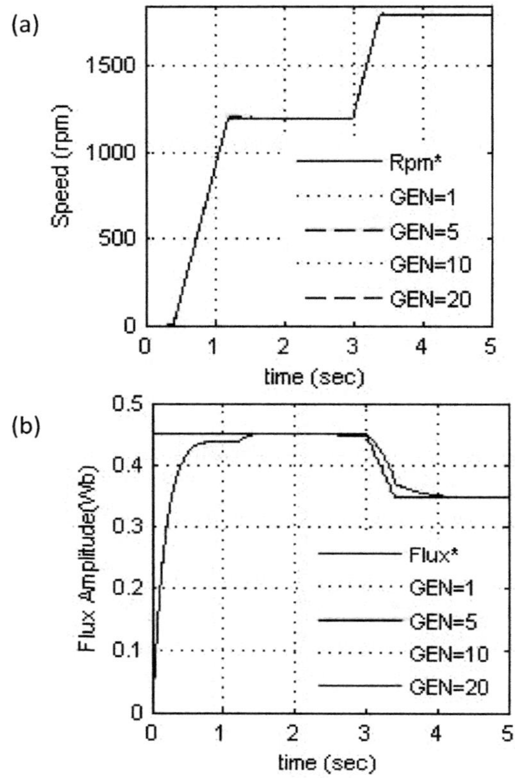

Figure 4.3 Speed tracking (a) and flux amplitude tracking (b) of indirect vector PI controller.

Table 4.3 Result of the fitness values after generation

	Fitness value		Speed		Current	
Generation	**Best (e-5)**	**Avg. (e-5)**	**K_p (e2)**	**K_i (e2)**	**K_p**	**K_i**
1	33.963	18.513	28.44	46.92	1.295	197.6
5	35.001	25.867	47.06	98.37	1.295	198.2
10	36.259	34.591	83.56	98.37	1.732	198.3
20	36.273	36.268	83.56	99.19	1.914	198.3

Figure 4.4a shows d-axis current and Fig. 4.4b illustrates q-axis current of indirect vector PI controller depending on generation of GA.

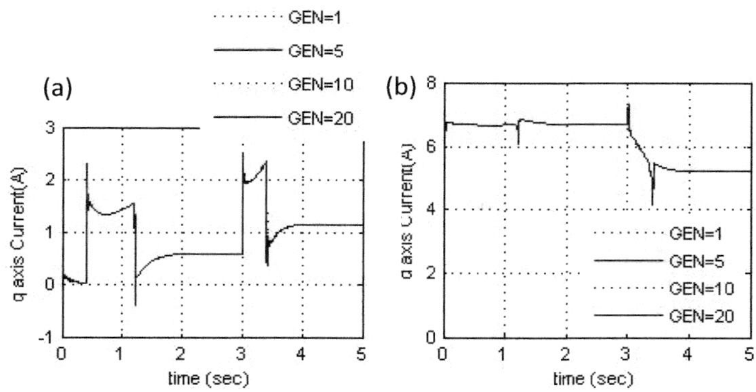

Figure 4.4 d-axis current (a) and q-axis current (b) of indirect vector PI controller.

4.7 Experiment and Discussion

In this section, TMS320F2812 DSP with 32 bit, 150 MIPS (TI Co.) and inverter is prepared for experiments.

TMS320F2812 is with Flash-ROM: 128 K × 16, SRAM: 18 K × 16, two SCI (communication port), and CAN function. Inverter is SV-iV5 with IGBT and voltage type (200–230 V, 50/60 Hz). Figure 4.5 shows the complete experimental setup.

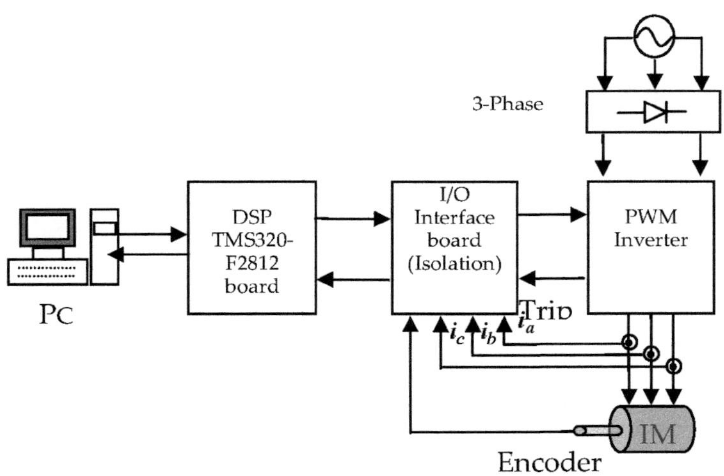

Figure 4.5 DSP-based indirect vector control system for induction machine.

Figures 4.6 and 4.7 show the results of simulation and experiments obtained by the optimal PI parameter. These figures represent speed tracking of GA-PSO-based indirect vector PI controller with generations 1 and 20 on experimental apparatus. The response curve is very similar to the speed tracking response obtained on simulation. Figures are *d*-axis current response of indirect vector PI controller.

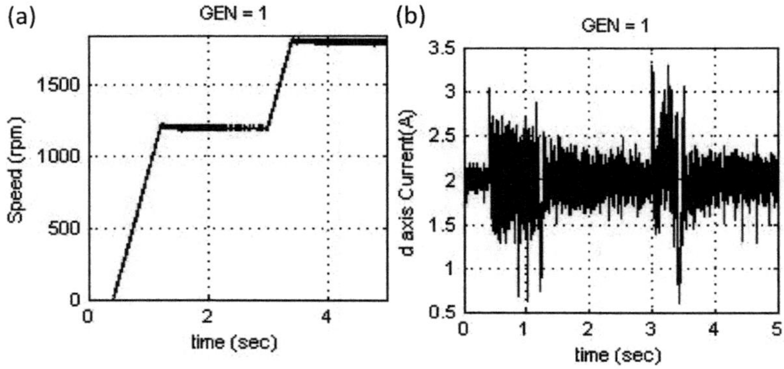

Figure 4.6 Speed tracking (a) and *d*-axis current tracking (b) in experiments of indirect vector PI controller (GEN = 1).

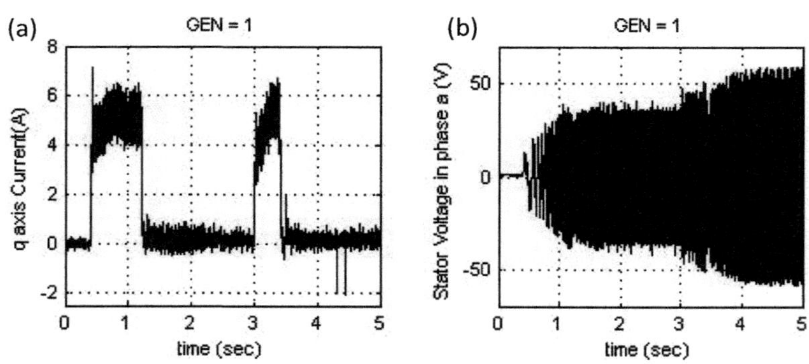

Figure 4.7 Experimental result of *q*-axis current (a) and stator voltage (b) in one phase of indirect vector PI controller (GEN = 1).

Figures 4.8a,b show the *d*-axis current tracking in experiments of indirect vector PI controller as GEN = 1 and GEN = 20.

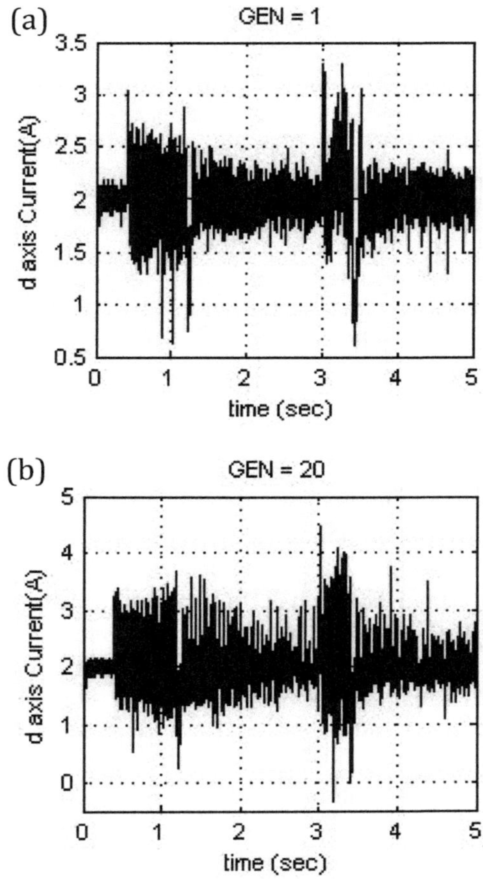

Figure 4.8 *d*-axis current tracking in experiments of indirect vector PI controller (GEN = 1, GEN = 20).

4.8 Summary

Even though several deep control theories have been developed, the proportional-integral-derivative (PID) controllers have been widely used owing to their simple structure, which can be easily understood and implemented for a wide range of process control, motor drives, flight control, and instrumentation. Owing to their popularity in the industrial world, over the past 50 years, several approaches for determining PID controller parameters have been developed for stable processes that are suitable for auto-tuning and adaptive control and for single input single output (SISO) systems. In spite of the enormous amount of research work reported in the tuning approaches, many PID controllers are poorly tuned in practice. One of the reasons is that most of the tuning methods are derived for particular processes and situations and therefore apply well only to their own areas. It is common experience that we are not certain which tuning method should be chosen to provide good control for a given process. Intelligent controllers could self-initialize and recalibrate even with little a priori knowledge especially due to the occurrence of significant changes in the process dynamics.

In this chapter, GA Euclidean-based distance concept and PSO were introduced for the enhancement of optimal tuning of conventional PID controllers for vector control of motor. By incorporating the Euclidean distance measure for selecting mutation or crossover points, the search space is well explored. Therefore, GA can provide exact optimal solution, while it can avoid local optimal solutions. On the other hand, this chapter also proposed a hybrid PSO approach to enhance the speed of convergence. A candidate solution for a specific problem in GA is called an individual or a chromosome and consists of a linear list of genes. Each individual represents a point in the search space, and hence a possible solution to the problem. A population consists of a finite number of individuals. Each individual is decided by an evaluating mechanism to obtain its fitness value. Based on this fitness value and undergoing GA-PSO operators with Euclidean distance, a new population is generated iteratively with each successive population referred to as a generation. Experiment results reveal the efficiency of the proposed approach with a faster convergence and optimal solutions. The GA-PSO system proposed in this chapter for PID controller tuning could be easily extended to model other complex problems involving local optimal and global optimal solutions.

Also, in this section, the GA-PSO-based indirect vector control is suggested through simulation and experiment. In optimal search by GA-PSO, a population consists of a finite number of individuals and each individual is decided by an evaluating mechanism to obtain its fitness value. Based on this fitness value and undergoing GA-PSO operators with Euclidian distance, a new population is generated iteratively with each successive population referred to as a generation.

The GA-PSO system proposed in this section for PI controller tuning of indirect vector control could be easily extended to model other complex problems involving local optimal and global optimal solutions.

Hence, a simultaneous tuning approach is evaluated in tuning of speed PI controller and current PI controllers for vector control of induction motor on simulation and experiments. Therefore, speed-following function torque, speed, and efficiency are able to be satisfied together. We expect this strategy for efficiency can be well performed for the load change.

4.9 Loss Minimization Control of Induction Motor Using GA-PSO

4.9.1 Overview for Loss Minimization

Induction motors have been widely using in various industries as actuators or drivers to produce mechanical motions and forces. Generally, variable-speed drives for induction motors require both wide operating range of speed and fast torque response in the required conditions, regardless of load variations. Since it is estimated that more than around 50% of the world electric energy generated is consumed by electric machines improving efficiency in electric drives is important [172, 173]. That is mainly, for two reasons as follows: economic saving and reduction of environmental pollution [174, 175]. Generally, since induction motors have a high efficiency at rated speed and torque, it has been widely used as driver in the industrial field. However, at light loads, iron losses increase dramatically, reducing considerably the efficiency [176, 177]. To improve the motor efficiency, the flux must be reduced, obtaining a balance between copper and iron losses [176]. Generally, there exist two different approaches to improve the efficiency in induction motor referred to as power measure–based method and loss model–based method:

In first approach, an optimum flux search algorithm is used, and the drive power consumption is measured [178–180]. This approach does not require the information or knowledge of motor parameters. However, it is only efficient at steady-state condition and for transients, a controller actuating as a supervisor, is needed to optimize efficiency with reduced flux levels. Different optimum flux search algorithms with an indirect field-oriented control are proposed and compared in a standard induction motor [3].

In second approach, when a motor loss model is available, the loss minimization optimum flux is computed analytically [175, 176]. The main advantage of this method is the simplicity of approach, not requiring extra hardware. However, it is compulsorily necessary to obtain an accurate knowledge of motor parameters, which change considerably with temperature, saturation, skin effect, variation of load, etc. In these methods, the optimum tools such as dynamic programming or intelligent optimization let us solve the problem of efficiency optimization to calculate the optimum flux trajectories. However, this approach presents difficulty to be implemented and it is also very sensitive to parameter changes.

To meet this requirement, many control approaches of induction motor are introduced into loss minimization control or energy saving control area. The field-oriented control is one of the most successful in meeting the above requirements. Due to advances in power electronics and DSP technology, variable-speed drives for induction motors using the field-oriented control have been widely used in many applications, such as AC servo, electric vehicle drive systems [172–176], and so on. Using the field-oriented control, a highly coupled, nonlinear, multivariable induction motor can be simply controlled by means of implementing linear independent decoupled control of torque and flux, similar to separately excited dc motors. Therefore, high-performance torque control requires fast enough current response for the current regulator to track the reference current. However, due to limitations of voltage and current ratings on the inverter dc link, input voltage and current of an induction motor are limited accordingly. Hence, developed torque in the motor should be limited for low loss operation under these input constraints.

In particular, since the objective of a variable-speed control system for higher productivity is to track the reference speed as fast as possible, under the constraints of input voltage and current, control approaches which yield the maximum torque over the entire speed range can be usefully applicable to minimum-time speed control of induction motors. However, most researchers who deal with the speed control of induction motors have not discussed the maximum-torque generation scheme, under low loss consideration.

In this section, a new minimum-time minimum-loss control algorithm for induction motors using hybrid system GA-PSO is suggested to obtain high performance, as well as high efficiency, under practical constraints on voltage and current. The validity of the suggested scheme, which carries out minimum-time speed control in the transient state and minimum-loss control in the steady state, will be revealed via simulation, including an induction motor model.

4.9.2 Mathematical Model of Induction Motor for Minimum Loss

Using the stator currents i_{ds}, i_{as}, and the rotor fluxes φ_{dr}, φ_{ar} as state variables, and by selecting a reference frame in such a way that the rotor flux, $\varphi_{ar} = 0$, the equations of the induction motor can be defined as [172, 186]

$$v_{ds} = R_s i_{ds} + \sigma l_s \frac{di_{ds}}{dt} + \frac{M}{l_r} + \frac{d\varphi_{dr}}{dt} - \omega\sigma l_s i_{ps} \tag{4.12}$$

$$v_{qs} = R_s i_{ds} + \sigma l_s \frac{di_{qs}}{dt} + \omega \frac{M}{l_r} v_{ds} i_{ds} + \omega\sigma l_s i_{ds} \tag{4.13}$$

$$M i_{qs} = \varphi_{dr} + \frac{l_r}{R} \frac{d\varphi_{dr}}{dt} \tag{4.14}$$

$$\omega = \omega_m + \frac{MR_r}{l_r} \frac{i_{qs}}{d\varphi} \tag{4.15}$$

$$T_e = \frac{M}{l_r} \varphi_{dr} i_{qs} = J\omega_m + KBJ\omega_m + T_r \tag{4.16}$$

$$\sigma = 1 - \frac{M^2}{l_r l_s} \tag{4.17}$$

In these equations, (R_s, L_s) and (R_r, L_r) are the stator and rotor equivalent windings resistance and inductance; M is the mutual inductance between rotor and stator dq windings; α is the leakage coefficient of the motor; ω_m and its derivative are, respectively the rotor angular speed and acceleration with respect to the stator; ω is the stator reference frame speed. The total loss in an induction motor can be defined as the difference between the input and output powers [178]:

$$\mu = \frac{P_{\text{out}}}{P_{\text{input}}} \quad (4.18)$$

$$P_{\text{input}} = v_{\text{ds}} i_{\text{ds}} + v_{\text{qs}} i_{\text{qs}} \quad (4.19)$$

$$P_{\text{out}} = \omega_r T_e, \quad (4.20)$$

where, ω_r is the rotor speed, T_e the electromagnetic torque, and v_{ds}, v_{ds}, i_{ds}, i_{qs} the stator voltages and the stator currents in the dq reference frame, respectively.

$$v_s = R_s i_s + \frac{d\varphi_s}{dt} + j\omega_s \varphi_s \quad (4.21)$$

4.9.3 Loss Minimization Vector Control Using GA-PSO

This section focuses on the application of the hybrid system GA-PSO described in the previous section for loss minimization control of an induction motor.

The proposed approach uses data points with the longest Euclidean distance for crossover process to avoid such local optimization. The idea is to obtain global solutions by considering the entire search space (all the data points) for loss minimization control. This section, initially, position of individuals is calculated by Euclidean distance–based method and then mutation and crossover are performed to improve the running speed and to obtain global optimal solution for loss minimization. Figure 4.9 is the vector control block diagram and Fig. 4.10 is the total block diagram of the control system based on GA-PSO for loss minimization control.

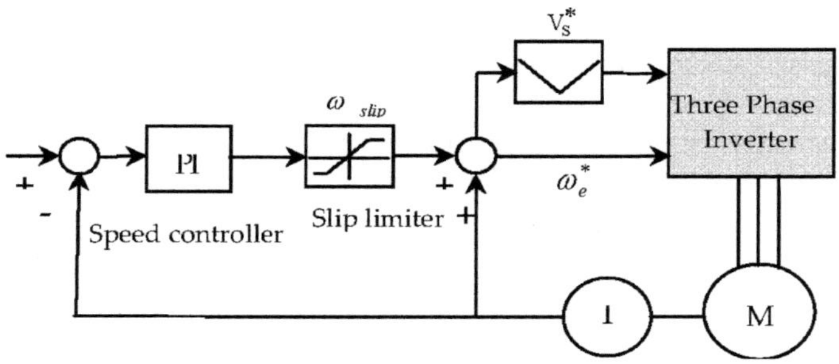

Figure 4.9 Vector control block diagram.

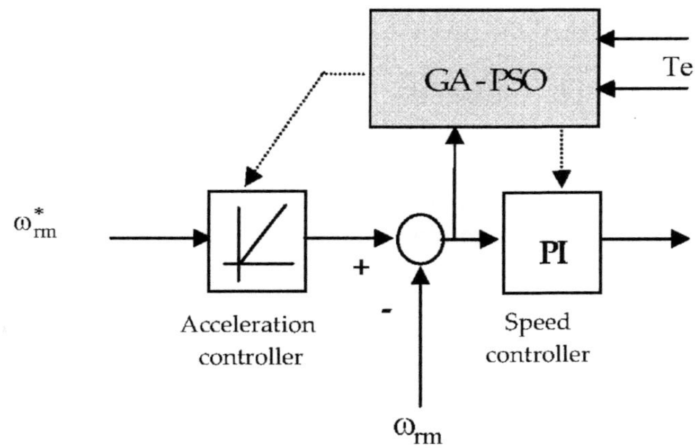

Figure 4.10 Control block diagram using GA-PSO.

In this section, fitness function of GA-PSO for performing loss minimization is defined as

$$\text{Fitness value} = \frac{1}{\alpha \cdot F_1 + \beta \cdot F_2 + F_c} \tag{4.22a}$$

$$F_1 = \frac{\sum_{i=0}^{N}(1-\eta(i))}{N}$$

$$F_2 = \frac{\sum_{i=0}^{N}\text{Speed error}}{N}, \tag{4.22b}$$

where F_e is the constant to avoid dividing by zero (≒1e−5), η efficiency, N the max time × sampling time, and α, β the weight constants.

Velocity-following error and efficiency of the motor are considered in Eq. (4.22), α and β are weighting functions.

Also, this section, as the control equation of acceleration speed for loss minimization operation, uses the following equation for acceleration time constant (ATC):

$$\text{ATC} = \frac{t \cdot \exp^{\gamma \cdot (T-t)}}{T}, \tag{4.23}$$

where T is the time taken to reach reference speed(s), γ the acceleration constant, and = T/Sampling time.

Figure 4.11 shows the speed characteristics curve of induction motor to variation of acceleration constant, γ. This section obtains the optimal tuning of PI controller using multi-objective function based on GA-PSO and acceleration constant γ for minimum-time and minimum-loss operation of induction motor.

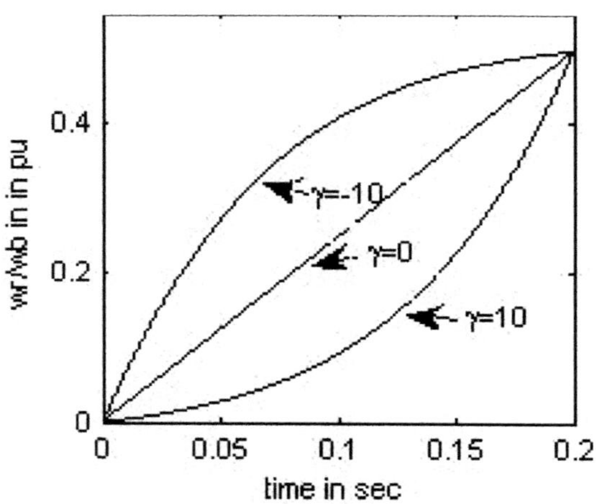

Figure 4.11 Speed characteristic to variation of acceleration constant, γ.

4.9.4 Simulation and Discussions

This section uses squirrel induction motor 20 HP, 220 V, 60 Hz with parameters as shown in Table 4.4 to simulate [174]. For safety, reference speed is defined as half speed of rated speed.

Weighting function of fitness function is arranged as ration of torque and speed (1:50).

Table 4.5 shows parameters of GA-PSO selected for motor simulation. Figure 4.9 show speed and torque curve to variation of rotor speed of induction motor used in this section.

Table 4.4 Parameters of induction motor for the simulation

Number of particles in a group	**4*3**
Maximum generation	30
Number of members in a particle	3 (P, I, γ)
Acceleration constant	2
Mutation rate	0.5
Maximum inertia weight factor	0.9
Minimum inertia weight factor	0.4

Table 4.5 GA-PSO Parameters

Parameter	**Symbol**	**Value**
Number of pole pairs	p	2
Stator resistance	r_s	0.1062 Ω
Referred rotor resistance	r_r'	0.0764 Ω
Stator magnetizing reactance	x_m	5.834 Ω
Stator leakage reactance	x_{ls}	0.2145 Ω
Rotor leakage reactance	x_{lr}'	0.2145 Ω
Rotor inertia in kgm^2	J_{rotor}	2.5 kgm^2

4.9.4.1 V/F control without PI current controller

In this section, to represent the characteristic of how volt/frequency device affect motor control system we simulate and illustrate the characteristic of PI speed controller shown in Fig. 4.9, we control motor using only volt/frequency control referred to in reference [175]. Figure 4.12 shows these characteristics. Speed performance is not good as expected. The speed function shown in Fig. 4.12a is stable after 0.2 s and the torque characteristic in Fig. 4.12b shows stable torque after 0.2 s. Figure 4.12b showing the torque characteristic without PI represents oscillation.

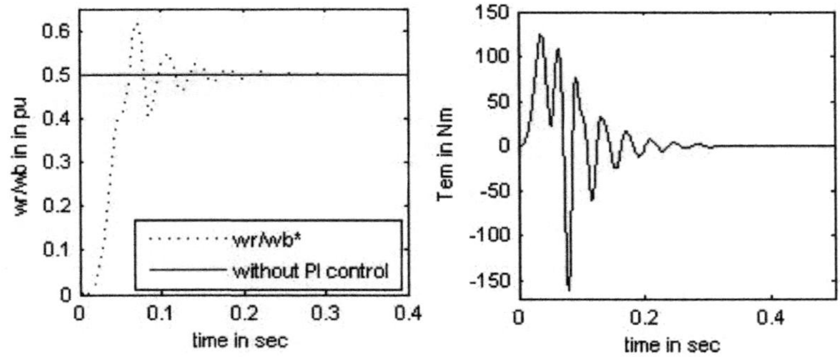

(a) Speed (b) Torque

Figure 4.12 Speed and torque characteristic without PI.

4.9.4.2 GA-PSO-based PI speed controller tuning

In this section, tuning of GA-PSO-based PI controller speed controller is performed and parameters are tuned by the GA-PSO illustrated in the previous section. As we have seen, the PI controller with parameters tuned by GA-PSO has much better speed and torque performance. That is, speed response shown in Fig. 4.13a by GA-PSO is faster and stable and the torque characteristic represents stable response within 0.2 s. However, more initial torque is required at the beginning of the start point. *P* and *I* parameters by operation after 30 generations are obtained as 9.918372 and 0.481056, respectively.

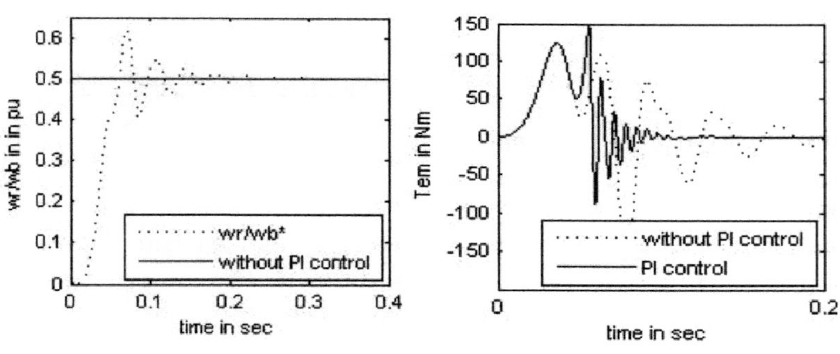

(a) Speed (b) Torque

Figure 4.13 Speed and torque following characteristics by GA-PSO-based tuning PI controller: GEN = 30.

4.9.4.3 Characteristics of variation of acceleration constant

In this section, the characteristic of the acceleration constant described in Eq. (4.23) is simulated. Since the efficiency of loss minimization control or energy saving control depends on the operation method such as the style of acceleration, we are going to introduce a moderate style of acceleration, as shown in Fig. 4.14, which depends on the variation and kinds of loads.

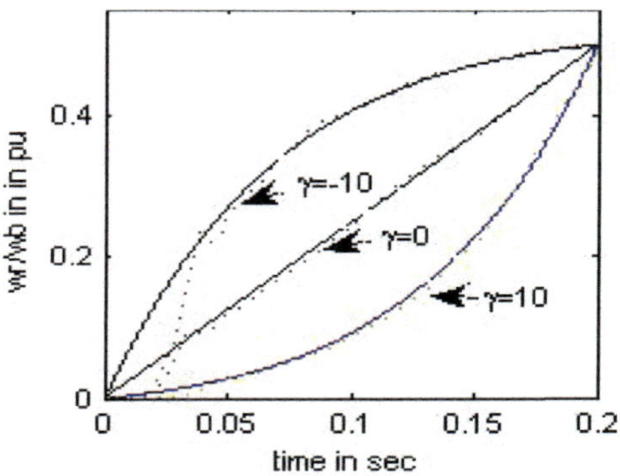

Figure 4.14 Speed curve to acceleration variation.

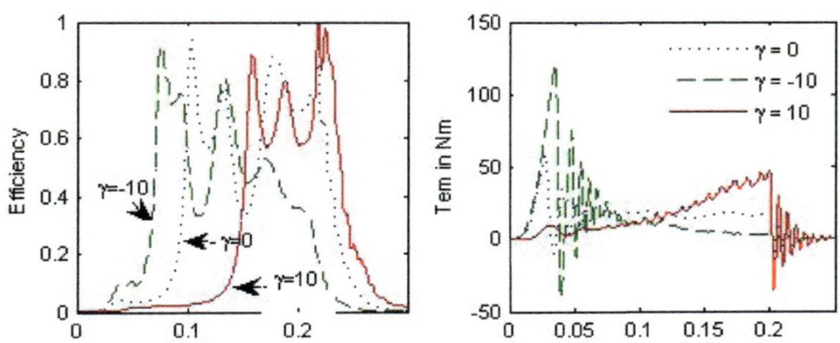

(a) Efficiency (b) Torque

Figure 4.15 Efficiency and torque curve to acceleration variation.

Figure 4.14 shows these characteristics. Figures 4.15a,b show the different efficiency and torque depending on the acceleration constant of Fig. 4.14. When PI parameters are fixed at the some value and γ is changed, the characteristic of induction motor is discussed. Figure 4.15 represents both characteristics such as efficiency and torque characteristics to acceleration variation. As we have seen in Fig. 4.14, efficiency and torque curves to acceleration variation shows much different shape depending on acceleration curves. Parameters tuned with GA-PSO mentioned in the previous section are used for this simulation and their values are 9.918372 and 0.481056, respectively.

4.9.4.4 GA-PSO-based tuning of PI parameters at fixed acceleration constant (T-PI-F-AC)

In this section, we tune PI parameters based on GA-PSO to the fixed acceleration constant γ. Figure 4.16 represents speed performance, torque characteristic, and efficiency curve to time when variations of generations are 10, 20, and 30, respectively. In torque of Fig. 4.17a and efficiency of Fig. 4.17b, results of GA-PSO-based control is lower torque at the starting point. That is, the Fig. 4.17a shows about 125 Nm as starting torque but shows around 55 Nm in the same condition in the previous method. Also, the efficiency curve of Fig. 4.17b shows that the results of GA-PSO-based tuning are more efficient.

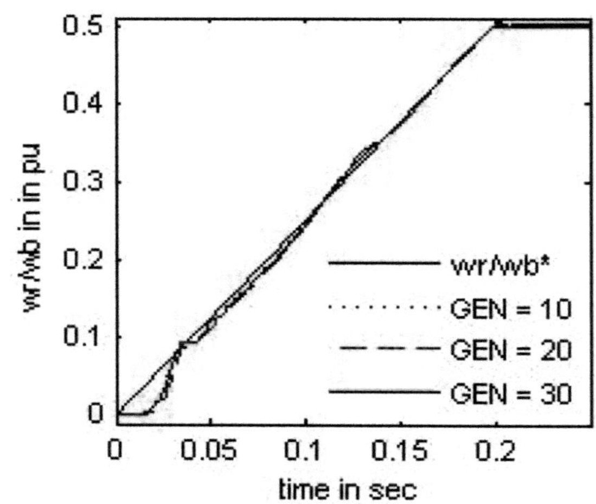

Figure 4.16 Speed performance of controller depending on generations (10, 20, 30).

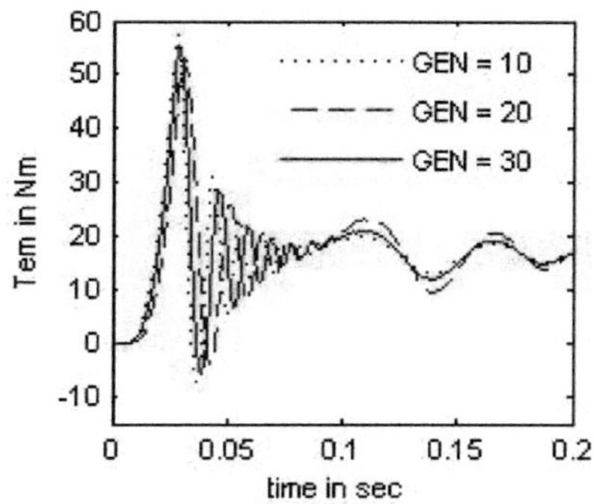

(a) Torque

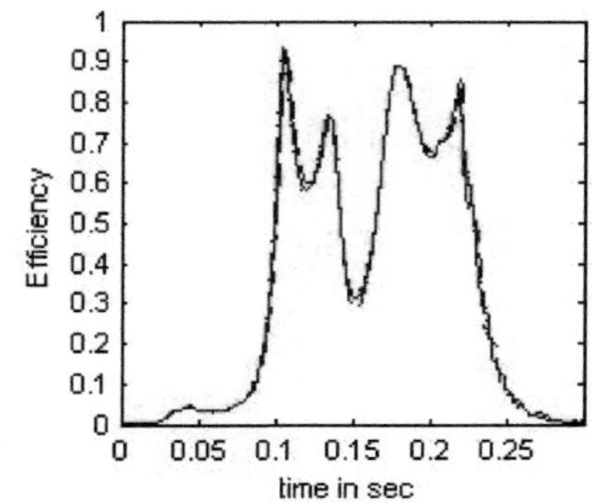

(b) Efficiency

Figure 4.17 Torque and efficiency characteristics of controller depending on generations (10, 20, 30).

Table 4.6 Parameters of generations (10, 20, 30)

Generation	*P*	*I*	Fitness value
10	3.6689	4.4939	0.9483
20	5.3141	2.9110	1.0490
30	6.8849	3.0689	1.1039

4.9.4.5 GA-PSO-based tuning of acceleration constant at fixed PI parameters (T-AC-F-PI)

Figure 4.18 shows speed, torque, and efficiency curves to acceleration constant γ tuned by GA-PSO at fixed PI parameter. Figure 4.18a shows that we can obtain moderate torque characteristic depending on load changing of control system, by GA-PSO-based acceleration constant γ tuning approach. The efficiency curve of Fig. 4.18a shows curve shape tuned by GA-PSO depending on the acceleration constant γ. This means that efficiency can be controlled to load changes by means of tuning acceleration constant γ by GA-PSO. In speed tracking curve of Fig. 4.18b, because of speed performance depending on acceleration constant γ tuned by GA-PSO, we can design good speed tracking controller by means of tuning of acceleration constant γ.

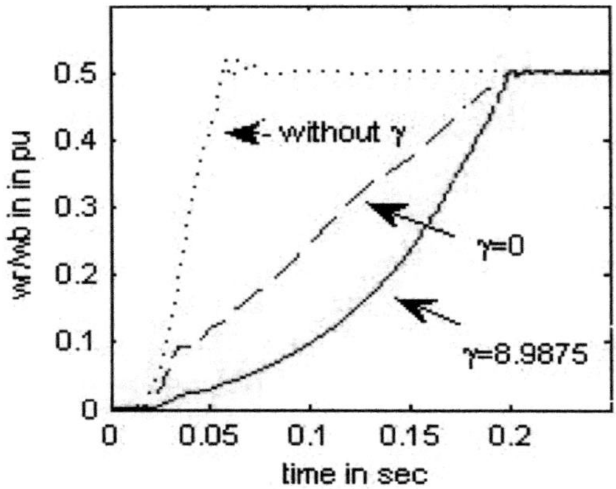

Figure 4.18 Speed tracking curve of PI Controller to acceleration constant γ tuned by GA-PSO.

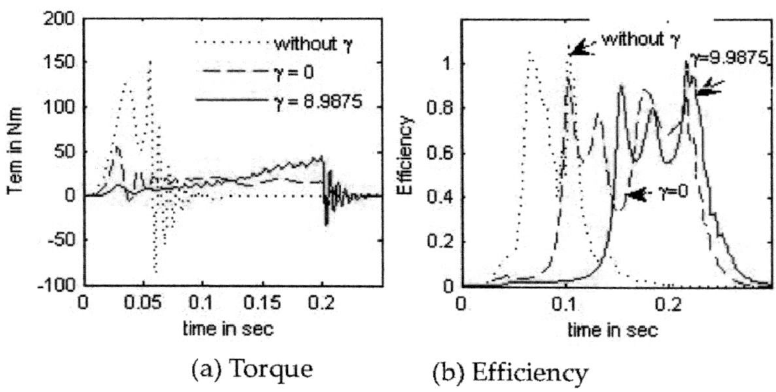

Figure 4.19 Torque and efficiency characteristics of PI Controller to acceleration constant γ tuned by GA-PSO.

4.9.4.6 Tuning of PI parameter and acceleration constant using GA-PSO (T-PI-T-AC)

This section tunes the acceleration constant γ and P and I parameters simultaneously using GA-PSO.

Figure 4.20 and Table 4.7 show the results of this simulation. Figure 4.20 illustrates the best characteristic acceleration curve is able to be obtained to the given load, by means of a simultaneous tuning of acceleration constant γ and PI parameter using GA-PSO.

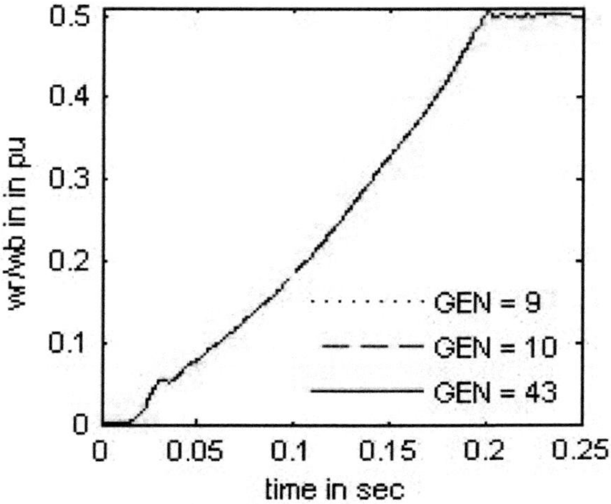

Figure 4.20 The acceleration curve characteristic to the given load, using the PI parameter based on GA-PSO at the tuned acceleration constant γ.

Also, Figs. 4.21a,b show the torque and efficiency characteristic curves using the PI parameter tuned by GA-PSO at the acceleration constant γ.

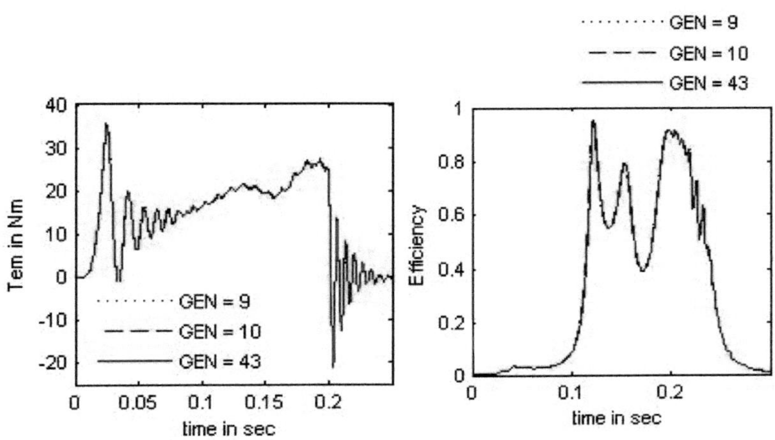

(a) Torque (b) Efficiency

Figure 4.21 Torque and efficiency characteristic using the PI parameter tuned by GA-PSO at acceleration constant γ.

Table 4.7 *P, I,* and *D* parameters tuned in generations

Generation	*P*	*I*	*D*	Fitness_value
5	9.8239	4.2431	2.7900	1.2050
10	9.9557	2.5413	2.7900	1.2074
43	9.9428	4.6268	2.9302	1.2078

4.10 Summary

In this section, the loss minimization optimum vector control is suggested. In this method, a simultaneous tuning approach of acceleration constant and PI parameters is used on GA-PSO. An optimal solution could be obtained by only a small number of individuals in GA-PSO, using multi-objective function of GA-PSO. As we can see in terms of dynamic responses mentioned in the previous section, T-PI-T-AC is as fast as T-PI-F-AC and T-AC-F-PI even with simultaneous tuning. T-AC-F-PI has a more complicated shape depending on variation of acceleration constant γ. T-PI-T-AC has the best behavior in terms of steady-state torque ripple, while T-AC-F-PI control shows larger ripple, but T-PI-F-AC is still much smaller than T-AC-F-PI. This strategy for efficiency can be well performed for the load change. Overall, the best strategy should be selected and performed depending on the load situation for real-world application.

Chapter 5

Intelligent Tuning Using Hybrid System of GA and BF

5.1 Background

In the past decade, evolutionary computation–based approaches have received increased attention from the engineers who deal with problems that cannot be solved using conventional problem-solving techniques [102, 103]. A typical task of a GA in this context is to find the best values of a predefined set of free parameters associated with either a process model or a control vector. One of the active areas of research in GA is system identification [93–95]. A recent survey of evolutionary algorithms for the evaluation of improved learning algorithms applied to control system engineering problems can be found in [13, 17, 19]. The general problem of evolutionary algorithm–based engineering system design has been tackled in various ways. GA has also been used to optimize nonlinear system strategies. Among them, a large amount of research is focused on the design of fuzzy controllers using evolutionary algorithm approaches. GA could be used for developing the knowledge base about the controlled process in the form of linguistic rules and the fine-tuning of fuzzy membership functions [22].

A possible solution to a specific problem can be encoded as an individual (or a chromosome), which consists of group of

Tuning Innovation with Biotechnology
Dong Hwa Kim

ISBN 978-981-4745-35-2 (Hardcover), 978-1-315-36458-2 (eBook)
www.panstanford.com

genes. Each individual represents a point in the search space and a possible solution to the problem can be formulated. A population consists of a finite number of individuals and the fate of each individual is decided by an evaluating mechanism that calculates its fitness value. Using this fitness value and genetic operators, a new population is iteratively generated referred to as a generation. The GA uses the basic reproduction operators such as crossover and mutation to produce the genetic composition of a population. The crossover operator produces two offspring's (new candidate solutions) by means of recombining the information from two parents.

As mutation operation is a random alteration of some gene values in an individual, the allele of each gene is a candidate for mutation, and its function is determined by the mutation probability. Many approaches to enhance the traditional GAs have been proposed [102, 103]. Among them, one category focuses on modifying the structure of the population or on the individual's role. Some examples are distributed GAs, cellular GA and symbiotic GAs. Another category is focused on modification/ efficient control of the basic operations, such as crossover or mutation, of traditional GAs [102–104].

On the other hand, as natural selection tends to eliminate animals with poor foraging strategies through methods for locating, handling, and ingesting food and favor the propagation of genes of those animals that have successful foraging strategies, such individuals are more likely to apply their reproductive success to find optimal solutions [115, 116]. After many generations, poor foraging strategies are either eliminated or shaped into good ones. Logically, such evolutionary principles have led scientists in the field of foraging theory to hypothesize that it is appropriate to model the foraging activity as an optimization process. Since a foraging animal takes actions to maximize the energy obtained per unit time spent foraging, in the face of constraints presented by its own physiology such as sensing, their cognitive capabilities, and the properties of environment (e.g., density of prey, risks from predators. physical characteristics of the search area), evolution can provide optimization within these constraints and essentially apply to some engineering fields what is sometimes referring to as an optimal foraging policy. That is, optimization models can provide for social foraging where groups of parameters communicate to cooperatively forage the modeled in engineering applications.

First, this section provides a brief literature overview of the area of bacterial foraging as it forms the biological foundation for this chapter. Then, this section also focuses on dealing with an enhanced optimal solution using a hybrid approach consisting of BF (Bacterial Foraging) and GA (Genetic Algorithm). Finally, we focus on analysis of the proposed hybrid systems by simulating using various test functions.

5.2 Hybrid System Consisting of GA and BF

As mentioned above, foraging behavior of bacteria [115–117] can be found, for instance using, dynamic programming. Search and optimal foraging-based decision-making of animals can be used to engineering applications. While performing social foraging an animal needs communication capabilities, it can gain advantages by essentially exploiting sensing capabilities of the group, the group can gang-up on large prey, individuals can obtain protection from predators while in a group, and in certain sense the group can forage with a type of collective intelligence.

5.2.1 Overview of Chemotactic Behavior of *Escherichia coli*

This section considers the foraging behavior of *E. coli*, which is a common type of bacteria [27, 28]. Its moving behavior comes from a set of up to six rigid 100–200 rps spinning flagella, each driven as a biological motor. An *E. coli* bacterium alternates through running and tumbling. Running speed is 10–20 μm/sec, but the bacteria cannot swim straight.

For this section, the chemotactic actions of bacteria can be summarized as follows:

- If in neutral medium, bacteria alternate by tumbles and runs, its action is exciting a search.
- If bacteria swims up a nutrient gradient (or out of noxious substances), it swims longer (climb up nutrient gradient or down noxious gradient); its behavior is to seek increasingly favorable environments.

- If bacteria swims down a nutrient gradient (or up noxious substance gradient), then the search action is to avoid unfavorable environments.

So, it can climb up nutrient hills and at the same time avoid noxious substances. The sensors it needs for optimal resolution are receptor proteins which are very sensitive and high gain. That is, a small change in the concentration of nutrients can cause a significant change in bacteria behavior. This is probably the best-understood sensory and decision-making system in biology.

Mutations in *E. coli* affect the reproductive efficiency at different temperatures, and occur at a rate of about 10^{-7} per gene and per generation. *E. coli* occasionally engages in a conjugation that affects the characteristics of a population of bacteria. Since there are many types of taxes that affect bacteria, such as: aerotaxis attracting to oxygen, phototaxis by light, thermotaxis by temperature, magnetotaxis, bacteria can be affected by magnetic lines of flux and some bacteria can change their shape and number of flagella which is based on the medium to reconfigure in order to ensure efficient foraging in a variety of media. Bacteria can form intricate stable spatio-temporal patterns in certain semisolid nutrient substances and they can eat radially their way through a medium if placed together initially at its center. Moreover, under certain conditions, they will secrete cell-to-cell attractant signals so that they will group and protect each other. These bacteria can swarm.

5.2.2 Optimization Function of Bacterial Swarm Foraging for the Hybrid System GA-BF

The main goal of applying the Hybrid GA-BF system based on bacterial foraging is to find the minimum of $P(\varphi)$, $\varphi \in R^n$, not in the gradient $\nabla P(\varphi)$. Here, φ is the position of a bacterium, and $J(\varphi)$ is the attractant-repellant profile. That is, it means where the nutrients and noxious substances are located, so the values $P < 0$, $P = 0$, $P > 0$ represent the presence of nutrients. A neutral medium, and the presence of noxious substances, respectively can be defined by [115, 116]

$$H(j,k,l) = \{\varphi^i(j,k,l) \,|\, i = 1, 2, \ldots, N\}. \tag{5.1}$$

The equation represents the positions of each member in the population of the N bacteria at the j-th chemotactic step, k-th reproduction step, and l-th elimination-dispersal event. Let $P(i, j, k, l)$ denote the cost at the location of the i-th bacterium $\varphi^i(j, k, l) \in R^n$, and

$$\varphi^i = (j+1, k, l) = \varphi^i(j,k,l) + C(i)\phi(j), \tag{5.2}$$

so that $C(i) > 0$ is the size of the step taken in the random direction specified by the tumble. If at $\varphi^i(j+1, k, l)$ the cost $J(i, j + 1, k, l)$ is better (lower) than at $\varphi^i(j, k, l)$, then another chemotactic step of size $C(i)$ in this same direction will be taken and repeated up to a maximum number of steps N_S. N_S is the length of the lifetime of the bacteria measured by the number of chemotactic steps. Functions $P_c^i(\varphi)$, $i = 1, 2, ..., S$, which model the cell-to-cell signaling via an attractant and a repellant are represented by [115–117]

$$\begin{aligned} P_c(\varphi) &= \sum_{i=1}^{N} P_{cc}^i \\ &= \sum_{i=1}^{N}\left[-L_{\text{attract}} \exp\left(-\delta_{\text{attract}} \sum_{j=1}^{n} (\varphi_j - \varphi_j^i)^2\right)\right] \\ &+ \sum_{i=1}^{N}\left[-K_{\text{repellant}} \exp\left(-\delta_{\text{repellant}} \sum_{j=1}^{n} (\varphi_j - \varphi_j^i)^2\right)\right], \end{aligned} \tag{5.3}$$

where $\varphi = [\varphi_1, ..., \varphi_p]^T$ is a point on the optimization domain, L_{attract} is the depth of the attractant released by the cell and is a measure of the width of the attractant signal. $K_{\text{repellant}} = L_{\text{attract}}$ is the height of the repellant effect magnitude, and δ_{attract} is a measure of the width of the repellant. The expression of $P_c(\varphi)$ means that its value does not depend on the nutrient concentration at position φ. That is, a bacterium with a high nutrient concentration secrets stronger attractant than one with a low nutrient concentration. The presented model uses function $P_{\text{ar}}(\varphi)$ to represent the environment-dependent cell-to-cell signaling as

$$P_{\text{ar}}(\varphi) = \exp(T - P(\varphi))P_c(\varphi), \tag{5.4}$$

where T is a tunable parameter. By minimizing the value of $P(i, j, k, l) + P_{ar}(\varphi^i(j, k, l))$ the cells try to find nutrients, avoid noxious substances, and at the same time try to move toward other cells, but not too close to them. The function $P_{ar}(\varphi^i(j, k, l))$ implies that, with M being constant, the smaller $P(\varphi)$, the larger $P_{ar}(\varphi)$ and thus the stronger attraction, which is intuitively reasonable. In tuning the parameter M, it is normally found that, when M is very large, $P_{ar}(\varphi)$ is much larger than $P(\varphi)$, and thus the profile of the search space is dominated by the chemical attractant secreted by *E. coli*.

On the other hand, if T is very small, then $P_{ar}(\varphi)$ is much smaller than $P(\varphi)$, and it is the effect of the nutrients that dominates. In $P_{ar}(\varphi)$, the scaling factor of $P_c(\varphi)$ is given as in exponential form. Below the method is presented in the form of an algorithm to search the optimal values of parameters.

Step 1: Initialize parameters n, N, N_c, N_s, N_{re}, N_{ed}, P_{ed}, $\alpha(i)$ $(I = 1, 2, ..., N)$, φ^i.

where

n: Dimension of the search space,

N: The number of bacteria in the population,

N_c: chemotactic steps,

N_s: the length of the lifetime in the bacteria,

N_{re}: The number of reproduction steps,

N_{ed}: the number of elimination-dispersal events,

P_{ed}: elimination-dispersal with probability,

$\alpha(i)$: the size of the step taken in the random direction specified by the tumble.

Step 2: Elimination-dispersal loop: $l = l + 1$

Step 3: Reproduction loop: $k = k + 1$

Step 4: Chemotaxis loop: $j = j + 1$

Substep a: For $i = 1, 2, ..., N$, take a chemotactic step for bacterium i as follows:

Substep b: Compute fitness function, ITSE (i, j, k, l).

Substep c: Let $\text{ITSE}_{last} = \text{ITSE}\ (i, j, k, l)$ to save this value since bacteria may find a better cost via a run.

Substep d: Tumble: generate a random vector $\theta_m(i) \in R^n$ with each element $\theta_m(i)$, $m = 1, 2, ..., p$, a random number in interval $[-1, 1]$.

Substep e: Move: Let

$$\varphi^i(j+1,k,l)=\varphi^i(j,k,l)+\alpha(i)\frac{\theta(i)}{\sqrt{\theta^T(i)\theta(i)}}.$$

This results in a step of size $\alpha(i)$ in the direction of the tumble for bacterium *i*.

Substep f: Compute ITSE $(i, j+1, k, l)$.

Substep g: Swim

(i) Let $m = 0$ (counter for swim length).

(ii) While $m < N_s$ (if have not climbed down too long).

- Let $m = m + 1$.
- If ITSE $(i, j+1, k, l) <$ ITSE$_{\text{last}}$ (if doing better), let ITSE$_{\text{last}}$ = ITSE $(i, j+1, k, l)$ and let

$$\varphi^i(j+1,k,l)=\varphi^i(j+1,k,l)+\alpha(i)\frac{\theta(i)}{\sqrt{\theta^T(i)\theta(i)}}$$

and use this $\varphi^i(j+1, k, l)$ to compute the new ITSE $(i, j+1, k, l)$ as did in

Substep h: Else, let m = N_S. This is the end of the while statement.

Substep i: Go to next bacterium $(i, 1)$ if $i \neq N$ (i.e., go to **[substep b]** to process the next bacterium).

Step 5: If $j < N_c$, go to step 3. In this case, continue chemotaxis, since the life of the bacteria is not over.

Step 6: Reproduction:

Substep a: For the given *k* and *l*, and for each $i = 1, 2, \ldots, N$, let

$$\text{ITSE}_{\text{health}}^i=\sum_{j=1}^{N_c+1}\text{ITSE}(i,j,k,l)$$

be the health of bacterium *i* (a measure of how many nutrients it got over its lifetime and how successful it was at avoiding noxious substances). Sort bacteria and chemotactic parameters $\alpha(i)$ in order of ascending cost ITSE$_{\text{health}}$ (higher cost means lower health).

Substep b: The S_r bacteria with the highest ITSE$_{\text{health}}$ values die and the other S_r bacteria with the best values split (this process is performed by the copies that are placed at the same location as their parents).

Step 7: If $k < N_{re}$, go to [step 3]. In this case, we have not reached the number of specified reproduction steps, so we start the next generation in the chemotactic loop.

Step 8: Elimination-dispersal: For $i = 1, 2, ..., N$, with probability P_{ed}, eliminate and disperse each bacterium, and this results in keeping the number of bacteria in the population constant.

To do this, if a bacterium is eliminated, simply disperse one to a random location on the optimization domain. If $l < N_{ed}$, then go to [step 2]; otherwise end.

5.3 Simulation for Comparison of the Conventional GA and the Hybrid GA-BF

This section illustrates the characteristics of the proposed hybrid GA-BF (genetic algorithms-bacteria foraging) and its relation to the conventional SGA (simple genetic algorithm), using the test function from De Jong [33].

5.3.1 Mutation Operation in GA-BF

Dynamic Mutation for illustrating the mutation characteristic in GA-BF is used here [33]. The mutation calculation is defined as [110–115]

$$\eta_j = \begin{cases} \eta_j + \theta(k, \eta_j^{(U)} - \eta_j), & \tau = 0 \\ \eta_j - \theta(k, \eta_j - \eta_j^{(L)}), & \tau = 1, \end{cases} \tag{5.4}$$

where random constant becomes 0 or 1 and is given as

$$\theta(k, y) = y \cdot \sigma \cdot \left(1 - \frac{k}{T}\right)^A, \tag{5.5}$$

where, σ has 0 or 1 randomly and T is maximum generations for computation. A is defined by user.

5.3.2 Crossover Operation in GA-BF

In this section, modified simple crossover is used to represent the BF-GA [33]. The chromosome in this approach is calculated as in [103]:

$$\eta_j^u = \lambda\eta_j^v + (1-\lambda)\eta_j^u, \tag{5.6a}$$

$$\eta_j^v = \lambda\eta_j^u + (1-\lambda)\eta_j^v, \tag{5.6b}$$

where η_j^u denote the generations, j the chromosome, and λ the multiplier.

5.3.3 Sensitivity to Variation of Step Size

For comparison of performance to variation of step size, the following test function from [104] is used:

$$F_1(x) = \sum_{i=1}^{3} x_i^2. \tag{5.7}$$

The test range is given as $-5.12 \le x_1, x_2, x_3\ 5.11 \le$. Step size means moving distance per step of the bacteria.

Figure 5.1 shows contour of test function F_1 to define shape of function before bacteria foraging.

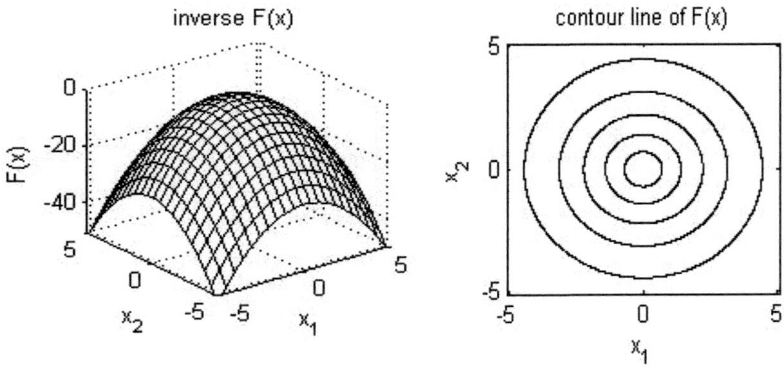

Figure 5.1 Contour of test function F_1.

5.3.4 Characteristics of Chemotactic Step of GA-BF

Figure 5.2 depicts characteristic to variation of step size when generations from 1 to 30 (Fig. 5.2a) and from 1 to 50 (Fig. 5.2b), respectively. As shown in Fig. 5.2, the bigger step size, the convergence is faster. Table 5.1 illustrates parameter values to step size of bacteria obtained by simulation on test function F_1.

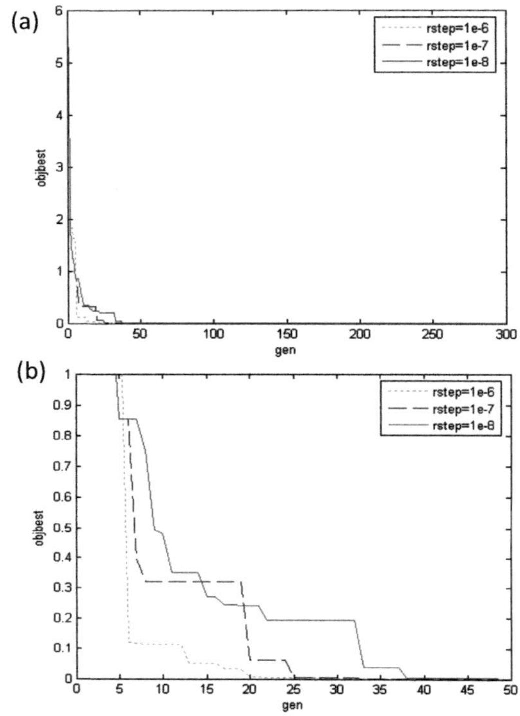

Figure 5.2 Characteristic of variables to variation of step size: (a) generations = 1–30, (b) generations = 1–50.

Table 5.1 Parameter values to step size of bacteria

Step size	x_1	x_2	x_3	Optimal objective function	Average objective function
1.0e-5	3.87E-13	6.60E-13	2.92E-07	-5.43E-07	-8.98E-08
1.0e-6	2.85E-14	2.34E-13	-5.52E-08	1.50E-07	-5.45E-08
1.0e-7	5.01E-16	1.43E-15	-1.70E-08	-1.44E-08	-2.31E-09

Figure 5.3 and Table 5.2 show the relationship between objective function and the number of generations in different chemotactic steps. When the chemotactic step is smaller, the objective function has a faster convergence with a small generator.

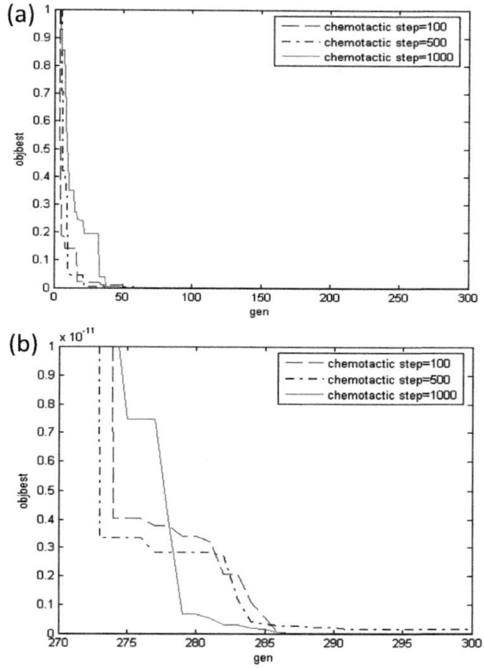

Figure 5.3 Relationship between objective function and generations in different chemotactic step: (a) generations = 0–300, (b) generations = 2700–300.

Table 5.2 Variation of parameter values to chemotactic step

Chemotactic step	x_1	x_2	x_3	Optimal objective function	Average objective function
100	−9.32E-08	3.78E-07	−8.57E-09	1.52E-13	1.59E-13
500	2.97E-08	1.92E-08	2.32E-08	1.79E-15	3.26E-15
1000	−1.70E-08	−1.44E-08	−2.31E-09	5.01E-16	1.43E-15

5.4 Characteristics of Life Time of GA-BF

Figure 5.4 shows characteristics between objective function and generators for different life time N_s of bacteria in the hybrid system, GA-BF.

Figure 5.4a is for objective function to generations 0 to 300 and Fig. 5.4b shows the results of generations 1 to 70. Figure 5.4b shows the shortest generations for obtaining objective function in $N_S = 1\text{–}4$.

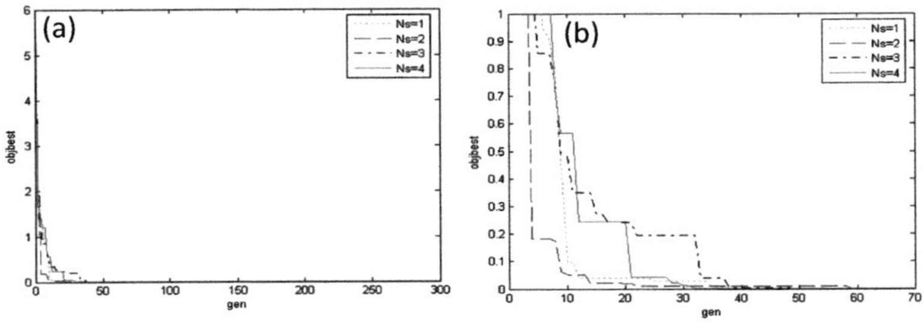

Figure 5.4 The characteristics between objective function and generators to different lifetime N_S of bacteria in the GA-BF: (a) generators = 0–300, (b) generators = 1–70.

Table 5.3 is an initial condition of the test function and variation of parameters obtained by simulation, and parameter values are obtained using range (searching range), chemotactic step, all number of chemotactic reaction of bacteria, step size, basic unit for movement of bacteria N_S, the number of critical reaction S, the number of bacteria G, generations Mu, mutation Cr, and crossover.

Table 5.3 Initial condition of test function and variation of parameters obtained by simulation

	Range		Genetic algorithm parameters				Bacteria foraging parameters			
Test function	$x_i^{(L)}$	$x_i^{(U)}$		**G**	**Mu**	**Cr**	**CS**	**Step size**	N_S	S
$F_1(x)=\sum_{i=1}^{3} x_i^2$	−5.12	5.11	20	300	0.9	0.1	1000	1e-007	3	10
$F_1(x)=100\,(x_1^2-x_2)^2+(1+x_1)^2$	−2.048	2.047	20	600	0.9	0.1	1000	1e-007	3	10
$F_2(x)=\sum_{i=1}^{5} x_i^2$	−5.12	5.12	20	180	0.9	0.1	1000	1e-007	3	10
$F_y(x)=\sum_{i=1}^{30} ix_i^4+N(0,1)$	−1.28	1.27	20	300	0.9	0.1	1000	1e-007	3	10

5.5 Characteristics of GA-BF to Test Functions

5.5.1 Test Function: f_1

Test function $f_1(x)=\sum_{i=1}^{3} x_i^2$ mentioned above (Fig. 5.1) is used for comparing the characteristic of the conventional GA and the proposed hybrid system GA-BF.

Figure 5.5a shows characteristics of GA and GA-BF on generation from 0 to 300 by this test function and Fig. 5.5b is characteristics of the hybrid GA and GA-BF on generations from 1 to 70. Figure 5.5b shows that the GA-BF can search optimal solution earlier as 10 generations as than the GA at the beginning step. Figure 5.6 shows the process of optimal search to test function in GA and GA-BF. Figures 5.7 and 5.8 illustrate the characteristics of GA and GA-BF with different step size (step size = 1×10^{-5}) by test function. Table 5.4 shows the parameters of GA and GA-BF obtained by simulation of the test function.

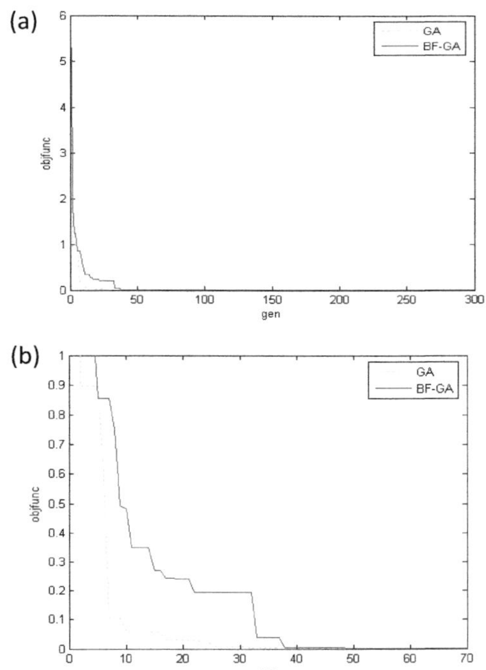

Figure 5.5 Characteristics of GA and GA-BF by test function f_1: (a) generators = 0–300, (b) generators = 1–70.

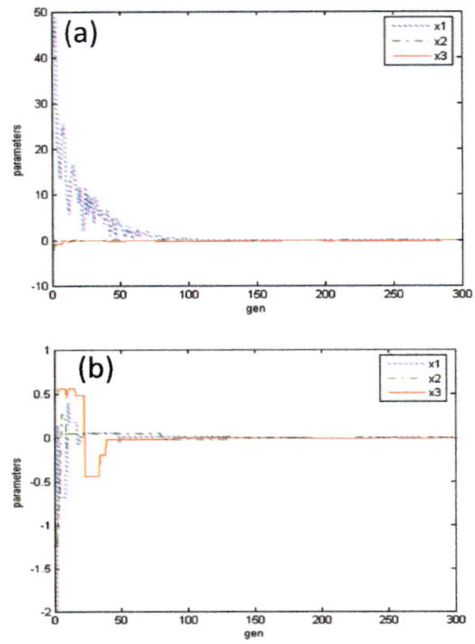

Figure 5.6 Process of optimal search using test function f_1: (a) GA, (b) GA-BF.

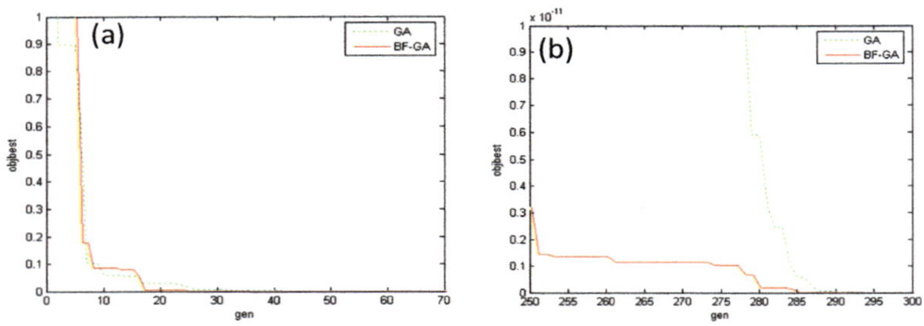

Figure 5.7 Characteristic of GA and GA-BF by test function f_1: Stepsize = 1 × 10^{-5}, (a) generation = 1–70, (b) generation: 270–300.

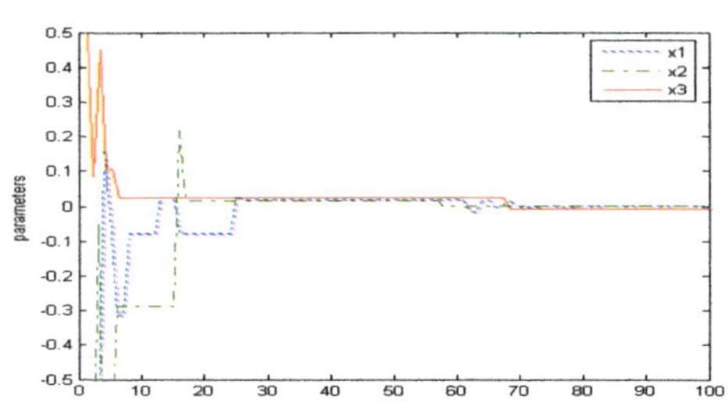

Figure 5.8 Process for optimal solution on GA and GA-BF using test function f_1: Stepsize = 1 × 10^{-5}.

Table 5.4 Parameter variation of GA and GA-BF by test function f_1

Chemotactic step	x_1	x_2	x_3	Optimal objective function	Average objective function
GA	7.22E−08	5.07E−08	−9.43E−09	7.87E−15	8.03E−15
GA-BF	−1.70E−08	−1.44E−08	−2.31E−09	5.01E−16	1.43E−15

5.5.2 Test Function: f_2

Test function $f_2(x) = 100(x_1^2 - x_2)^2 + (1 - x_1)^2$ is used for comparison of performance of GA and GA-FA.

Figure 5.9 shows the contour of this function at $x = [11]^T$. Figures 5.10 and 5.11 represent characteristics of the conventional GA and the proposed hybrid system GA-BF for the test function (generation: 1–70), as the previous test function.

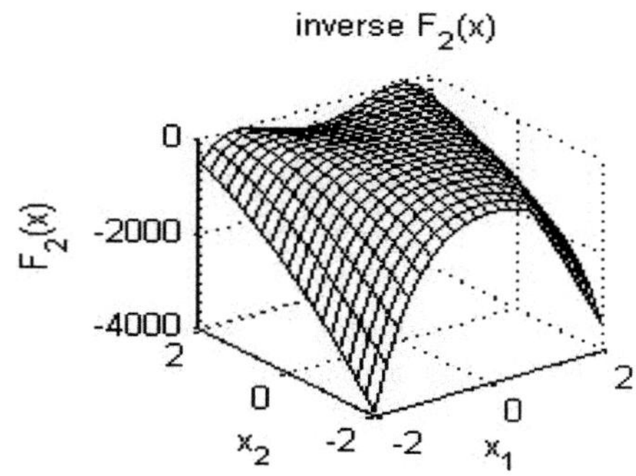

Figure 5.9 Contour of test function (F_2) at $x = [11]^T$.

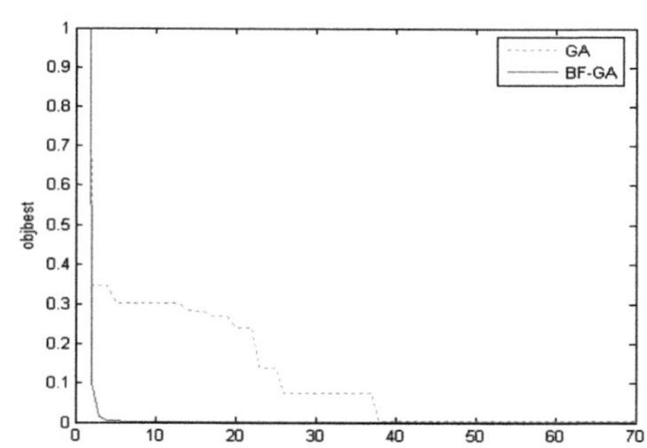

Figure 5.10 Characteristic of GA and GA-BF by test function f_2: generations = 1–70.

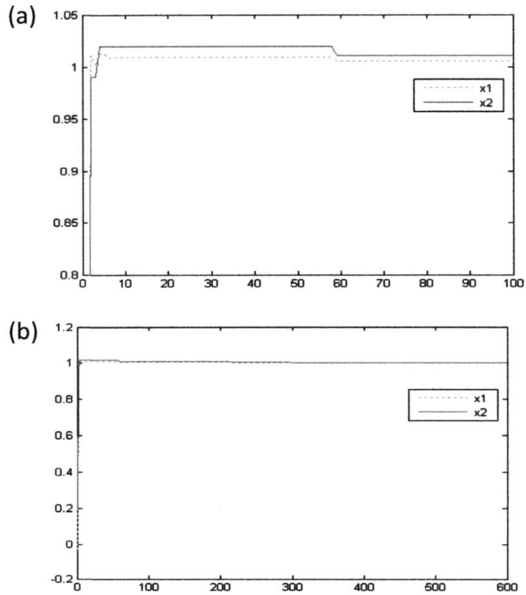

Figure 5.11 Optimal search process of GA (a) and GA-BF (b) in test function, f_2.

From these figures, GA-BF can find the solution earlier than GA at all generations (initial generations and final generations) and we can see that it is possible for the GA-BF to find a more satisfactory solution. Table 5.5 includes parameter values of GA and GA-BF obtained for test function.

Table 5.5 Parameter variation of GA and GA-BF by test function f_2

Chemotactic step	x_1	x_2	Optimal objective value	Average objective value
GA	0.001967	0.001967	1.0443267	1.0907699
BF-GA	5.12E-09	5.17E-09	0.9999285	0.9998567

5.5.3 Test Function: f_3

Test function $f_3 = \sum_{i=1}^{5} [x_i]$ is applied to compare performance of GA-BF and GA. This function has the minimum of −30 at

x = [−5.12, −5.12, −5.12, −5.12, −5.12]. Figure 5.12 is contour map of this function and Figs. 5.13 and 5.14 represent the result simulated by this test function as the test function.

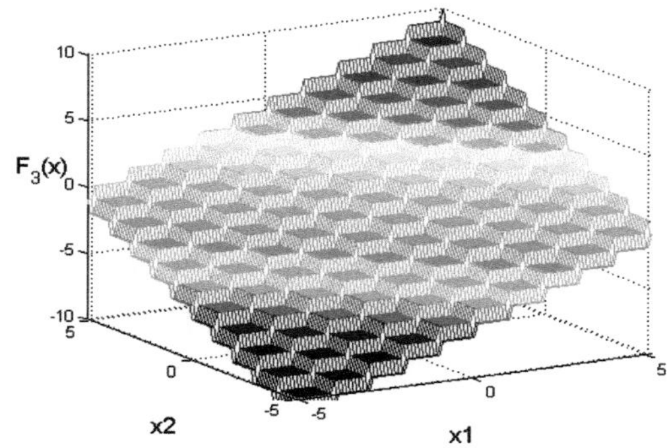

Figure 5.12 Contour map of test function f_3.

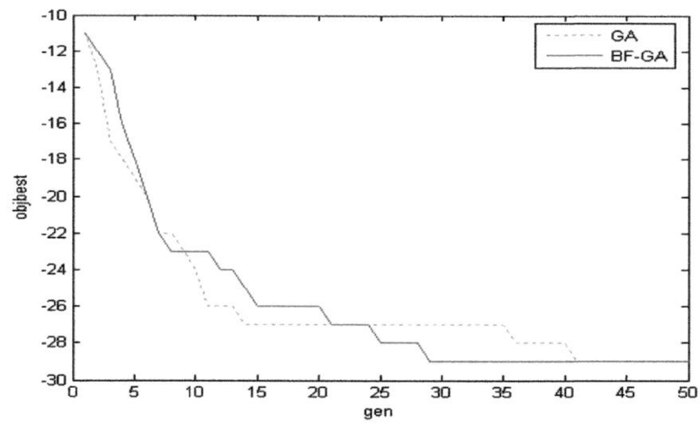

Figure 5.13 Characteristic of GA and GA-BF by test function f_3: generations = 1–70.

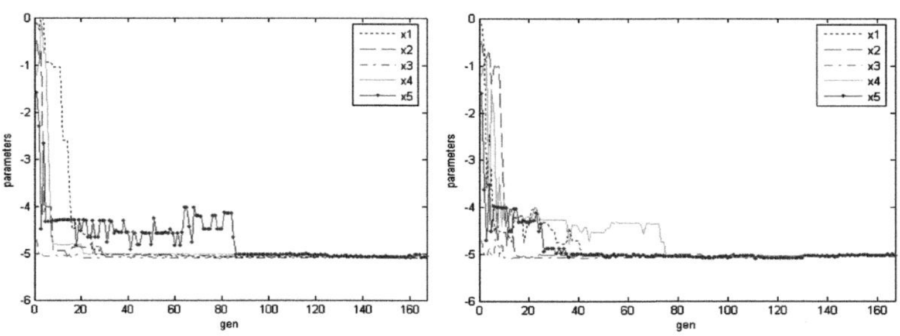

Figure 5.14 Optimal search process of GA (a) and GA-BF (b) on test function f_3.

Table 5.6 Parameter variation of GA and GA-BF by test function f_3.

Method	x_1	x_2	x_3	x_4	x_5	Optimal objective value	Average objective value
GA	-5.024811	-5.015523	-5.059941	-5.03529	-5.03527	-30	-29.4
BF-GA	-5.111186	-5.097807	-5.089435	-5.06529	-5.06891	-30	-29.95

5.5.4 Test Function: f_4

$f_4 = \sum_{i=1}^{30} ix_i^4 + N(0,1)$ is used to compare the conventional GA and the proposed system GA-BF and Fig. 5.15 shows the contour map of this function. In Fig. 5.16, GA and GA-BF represent a similar response, but Fig. 5.16 shows the hybrid system BF-GA is faster.

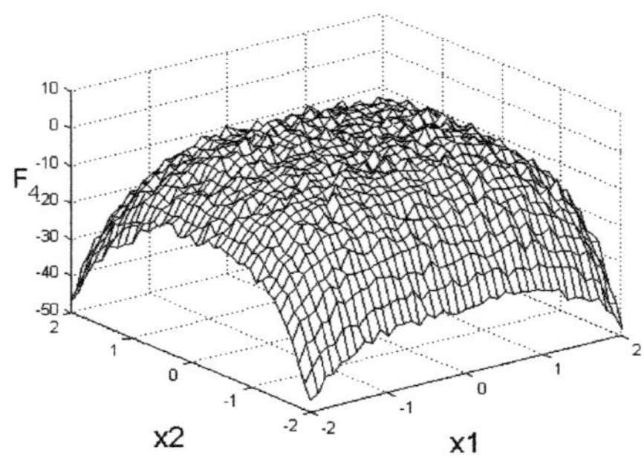

Figure 5.15 Contour map of test function f_4.

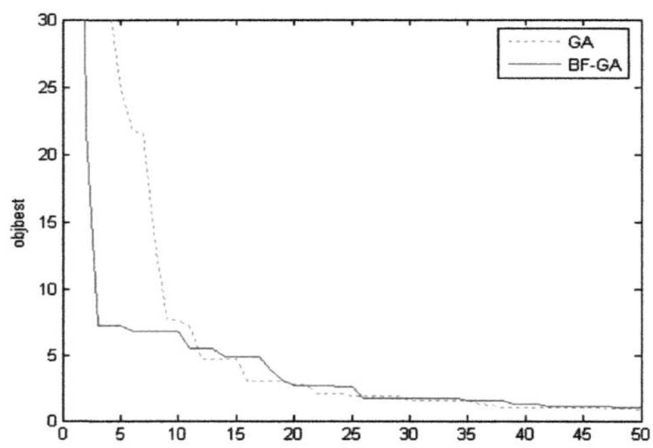

Figure 5.16 Characteristic of GA and GA-BF by test function f_4: generations = 1–70.

Figure 5.17 represents optimal search process of GA (a) and GA-BF (b) on test function f_4.

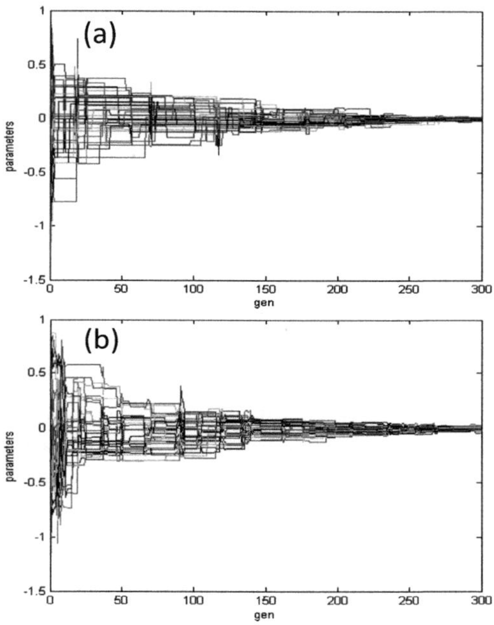

Figure 5.17 Optimal search process of GA (a) and GA-BF (b) on test function f_4.

Table 5.7 Parameter variation of GA and GA-BF by test function f_4

Method	x_1	x_2	x_3	x_4	x_5	Optimal objective value	Average objective value
GA	−5.024811	−5.015523	−5.059941	−5.03529	−5.03527	−30	−29.4
BF-GA	−5.111186	−5.097807	−5.089435	−5.06529	−5.06891	−30	−29.95

5.6 Intelligent Tuning of AVR System Using GA-BF

The GA-BF algorithm suggested in this section is used in PID Controller tuning of AVR system mentioned earlier.

The initial values of the PID Controller and GA-BF for simulation are shown in Tables 5.8 and 5.9, respectively.

Table 5.8 Range of PID parameters for learning of GA-BF

Parameter	Range	
	Min	Max
k_p	0	1.5
k_i	0	1
k_d	0	1

Table 5.9 Simulation parameters of GA-BF

Parameters	Values
Step size	0.0
N_s	4
P_c	0.9
P_m	0.65

Figure 5.18 shows terminal voltage step response of an AVR system with GA-BF-based tuning controller to weighting factor β = 1.0 and β = 1.5. It shows the results obtained by GA, PSO, GA-PSO, and GA-BF (BF-GA) to variation of β = 0.5 and β = 1.5 at generations 200.

(a)
(b)
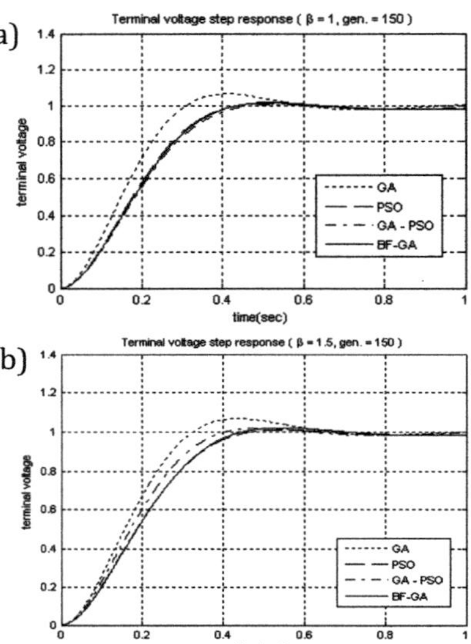

Figure 5.18 Terminal voltage step response of an AVR system with different controllers (weighting factor β = 1.0 (a), β = 1.5 (b), generations = 200).

Results of GA-BF show best satisfactory learning.

Figure 5.19 is search process for optimal parameters in an AVR system by GA-BF (β = 1.5).

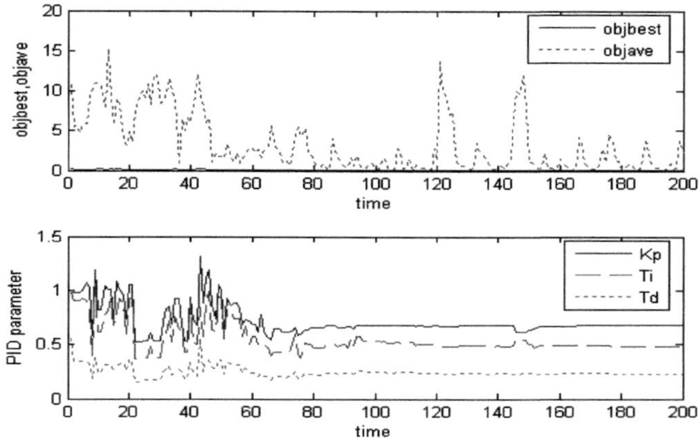

Figure 5.19 Search process for optimal parameters in an AVR system by GA-BF (β = 1.5).

Table 5.10 shows the best solution using GA-BF controller with the different values and Table 5.11 shows the comparison of the value evaluated by both methods (β = 1.5, generation = 200).

Table 5.10 Best solution using GA-BF controller with the different β values

β	No. of gen.	k_p	k_i	k_d	Mo (%)	ess	t_s	t_r
0.5	200	0.68233	0.6138	0.26782	1.94	0.0171	0.3770	0.2522
1	200	0.68002	0.52212	0.24401	1.97	0.0067	0.4010	0.2684
1.5	200	0.67278	0.47869	0.22987	1.97	0.0014	0.4180	0.2795

Table 5.11 Comparison of the evaluation value between both methods (weight function β = 1.5, generation = 200)

β	No. of gen.	k_p	k_i	k_d	Mo (%)	ess	t_s	t_r
1.5	GA	0.8282	0.7143	0.3010	6.7122	0.0112	0.5950	0.2156
	PSO	0.6445	0.5043	0.2348	0.8399	0.0084	0.4300	0.2827
	GA-PSO	0.6794	0.6167	0.2681	1.8540	0.0178	0.8000	0.2526
	BF-GA	0.6728	0.4787	0.2299	1.97	0.0014	0.4180	0.2795

5.7 GA-BF-Based PI Controller Tuning for Indirect Vector Control of Three Phase Induction Motor

In this section, GA-BF-based PI controller is suggested for the tuning of indirect vector control of three-phase induction motor mentioned in Section 3.5. The overview and structure of the vector control for the induction motor are similar to the structure explained in Section 3.5.

Here, the GA-BF-based PI controller is used for vector control. Figure 5.20 illustrates speed tracking in the induction motor by simulation (Fig. 5.20a) and by the experiment (Fig. 5.20b). The shapes of these figures are almost similar.

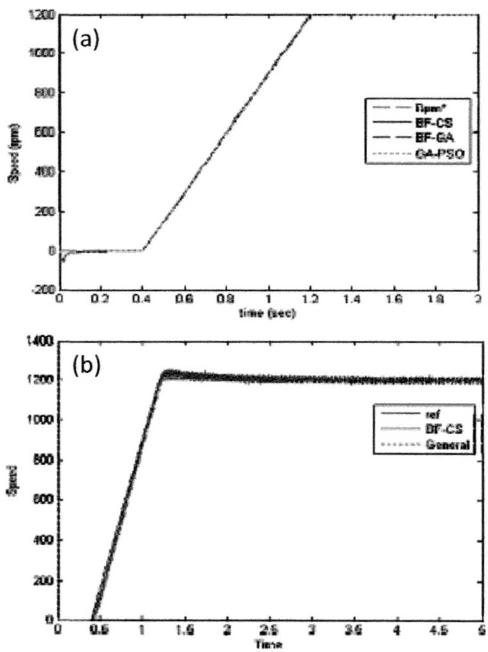

Figure 5.20 Speed tracking in induction motor by each approach (a) simulation, (b) experiment.

Figure 5.21a shows torque tracking by experiment and Fig. 5.21b shows the d-axis and q-axis tracking in the induction motor by the experiment.

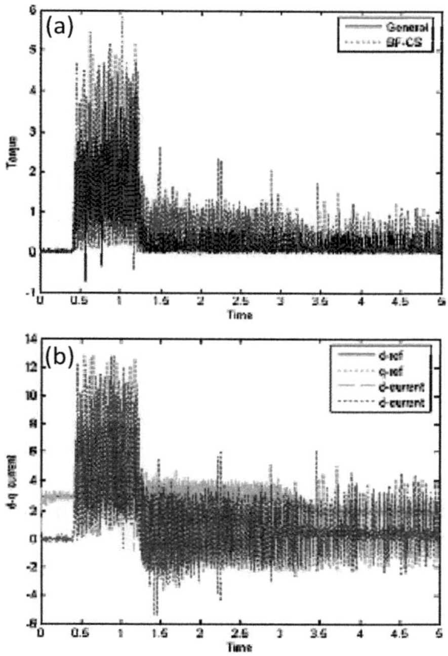

Figure 5.21 Torque tracking (a) and d-axis and q-axis tracking (b) in induction motor by experiment.

Table 5.12 gives objective function, parameters of speed PI controller, and current PI controller tuned by GA-BF and GA PSO.

Table 5.13 provides ISE (integral square error) in reference [1] and used in GA-BF.

Table 5.12 Objective function and parameters in each approach

	Obj. Fn.	Speed		Current	
Method	Best (10^{-5})	K_p (10^{-2})	K_i (10^{-2})	K_p	K_i
BF-GA	28.263	0.99896	0.55318	4.5462	82.794
GA-PSO	36.273	83.56	99.19	1.914	198.3
BF-CS	31.345	1	1	13.657	175.63

Table 5.13 ISE (integral square error) in each method

Method	ISE (integral square error) in speed
Ref. [1]	1.7371×10^{-6}
GA-BF	1.4251×10^{-6}

5.8 Summary

Several recent approaches of evolutionary algorithms for the evaluation of improved learning algorithm and control engineering have been studied. The general problem of evolutionary algorithm-based engineering system design has been tackled in various ways because of the learning time and local or suboptimal solution.

GA has also been used to optimize nonlinear system strategies, but it might be local-optimized. On optimization, vast research is focused on the design of fuzzy controllers using evolutionary algorithm approaches. GA could be used for developing the knowledge-based learning about the controlled process in the form of linguistic rules and the fine-tuning of fuzzy membership function. However, it may also have problem with local optimization or suboptimal solution.

This chapter proposes the hybrid system consisting of GA (genetic algorithm) and BF (bacterial foraging) and proves the characteristic of the system using various test functions.

This approach proposed in this chapter has the potential to be useful in practical optimization problems (e.g., engineering design, online distributed optimization in distributed computing and cooperative control) as models of social foraging are also distributed nongradient optimization methods. It may also be used for a wide variety of fruitful research directions and ways to improve the models (e.g., modeling more dynamics of cell motion).

Moreover, other species of bacteria or biological-based computing approach could be studied, but it remains to be seen how practically useful the optimization algorithms are for engineering optimization problems, because they depend on the theoretical properties of the algorithm, theoretical and empirical comparisons to other methods, and extensive evaluation on many benchmark problems and real-world problems.

In this chapter, we introduce the suggested approach into PID controller tuning of AVR system and tuning of PI controller for the vector control of the induction motor. Each approach shows satisfactory results.

Chapter 6

Artificial Intelligence, Emotion Function, and ICT

6.1 Overview

This chapter explains the research motivation for bio-based tuning issue such as particle swarm optimization, immune algorithm, bacterial foraging, and its hybrid system for tuning. In this chapter, the emotion control approach of PID controller with disturbance acceptance function against external emotion in control system is proposed using bacterial foraging–based optimal algorithm. Up to the present time, the PID controller has been used to operate for disturbance reject because disturbance is not used in practice structure. However, this chapter deals with disturbance positively for emotion control. Also, to obtain optimal control gain, this chapter uses the bacterial foraging algorithm. This chapter illustrates disturbance acceptance conditions and (integral of time weighted squared error (ITSE) as the performance index. The parameters of PID controller are selected by bacterial foraging–based optimal algorithm to obtain the required response for emotion control.

Section 6.3 shows emotion control on induction motor. Emotions have always appeared to be a complex phenomenon and many researchers have been interested in express or recognition. However, it is not easy for an engineer to produce a user-

Tuning Innovation with Biotechnology
Dong Hwa Kim

ISBN 978-981-4745-35-2 (Hardcover), 978-1-315-36458-2 (eBook)
www.panstanford.com

friendly emotion function because emotion is very complicated in expression. We have tried to illustrate emotions on fuzzy agent and control such as positive and negative functions in stimulation and suppression. Stimulations, suppressions, and decay function is important to emotion resource. This chapter discusses the work done on the above and on the three emotions generated from percepts gathered by a robot moving along a path.

The emotion function of the human mind has an important role in decision-making, memory, action, and good communication or so. Especially, the emotion characteristics of voice are very important for warm communication, successful business, human-to-human good relationship, and good care for children and silver ages. On the other hand, recently, the service robot market for the educator, helper, secretary, deliverer, and guide has been growing up because of old population and complicated social situation. In that case, the emotion function is needed in those areas.

Section 6.4 describes the application of emotion function to voice control. The emotion characteristic of voice depends on pitch contour, acoustic energy, vocal tract features, speech energy, etc. Therefore, we need to consider how we have to apply and implement emotion function of voice for a service robot. However, its implementation for a robot is very difficult and recognition is also not easy because of various emotion patterns in voice. This chapter suggests a method of voice emotion generation for user demand emotion talk in service robot. Fuzzy rule–based approach is introduced to generate emotion for user demand emotion function by controlling pitch contour, acoustic energy, vocal tract features, and speech energy.

6.2 The Design of 3-DOF Emotion Controller

Emotion technology and PID controller are very useful. However, emotion technology faces some difficulty in implementation in the actual system. But PID control has been used from long time ago. The proportional-integral-derivative (PID) controllers have been widely used owing to their simple structure. In industrial applications, more than 90% of all control systems are PID type [6, 174–178]. Also, owing to their popularity in the industrial world, several approaches for determining PID controller parameters have been developed for tuning [6, 25, 174–177]. These methods are straightforward to apply since they provide simple tuning formulas to determine the PID controller parameters. On the other hand, 2-DOF PID controller has been suggested for recovering disturbance and set-point together [9, 10] and its tuning approaches are also studied in many methods. On the other hand, it is very important to express the human being mind, which has an important role in decision-making [9]. Therefore, emotion is a popular topic in AI research, but most of existing work focuses on the appraisal of emotions and mechanisms eliciting emotions. Primary emotions can rise from external stimulus patterns. They can result in fast, species-typical reactions to environmental changes. Therefore, the emotion technology should be studied. It will be used in IT technology and design. Neural, neuro-fuzzy, and evolutionary systems and their derivatives followed later. Biological information processing systems such as human beings have many interesting functions and are expected to provide various feasible ideas to engineering fields [10]. Biological information such as immune system in living organisms can be mainly classified into the following four systems: brain, nervous system, genetic system, and endocrine system. Also, they are the important for artificial intelligence. This chapter suggests the scheme and tuning approach for the 3-DOF Emotion Intelligent PID controller (3-DOF-EI-PID).

This type of the 3-DOF-EI-PID controller has a combined parameter for a 3-EI-DOF function as shown in Fig. 6.1. The transfer functions between process value PV(s) and settling value SV(s) and between process value PV(s) and disturbance DV(s) are given in the following equations:

$$G_{\text{PVDV}}(s) = \frac{PV(s)}{DV(s)} = \frac{G_d(s)}{1 + K_p\left(1 + \frac{1}{T_i s}\right)\gamma} \tag{6.1}$$

$$G_{\text{PVSV}}(S) = \frac{PV(s)}{SV(s)} = \frac{\alpha K_p\left(1 + \frac{1}{T_i s}\right)\left(\frac{1}{1 + \beta T_i s}\right) + \left(\frac{\beta K_p K_d s}{1 + \eta T_d s}\right)}{1 + K_p\left(1 + \frac{1}{T_i s}\right)\gamma} SV \tag{6.2}$$

$$G_{\text{PVG1}}(s) = \frac{\left(\dfrac{\beta K_p K_d s}{1+\eta T_d s}\right)}{1+K_p\left(1+\dfrac{1}{T_i s}\right)\gamma} G_1(s), \tag{6.3}$$

where, the filter transfer function is $F(s) = 1/(1 + \beta T_i s)$, the PI controller transfer function is $\text{PI}(s) = K_p/(1 + 1T_i s)$, and the D controller transfer function is $D(s) = (K_p T_d(s)/(1 + 1\sigma T_d s)$. In Eq. (6.1), the numerator has a function similar to that of the conventional PID controller.

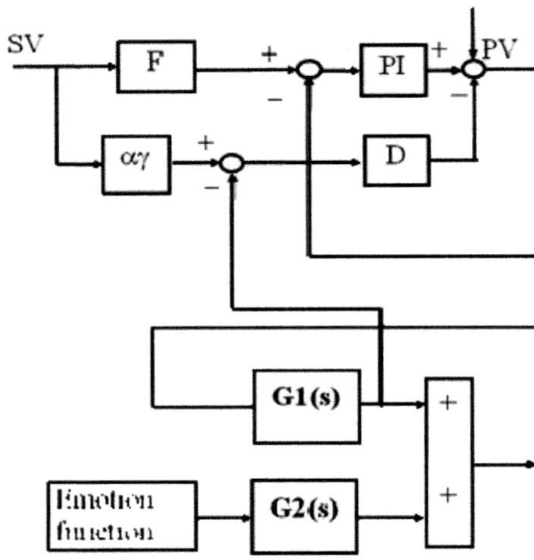

Figure 6.1 Block diagram of the design principle of the 3-DOF PID controller with an emotion function.

That is, if the proportional gain K_p goes to a greater value, the efficiency of disturbance G_d is smaller. However, in Eqs. (6.2) and (6.3), the process value PV(s) and the plant $G_1(s)$ depend on the two degrees of parameter α, β, γ. The proportional gain could also be affected by the parameter α, β, and γ given for the two degrees function. Since the disturbance can be reduced by gains K_p, T_i, and γ the process value PV and the plant $G_1(s)$ are effectively controlled by the two degrees parameters, α, β, γ. Then, a 3-DOF-EI-PID controller can perform the two degrees of function, completely. The result of this arrangement distinguishes it from the conventional arrangement method.

Trigger level of emotion function depends on the emotion function (Fig. 6.2).

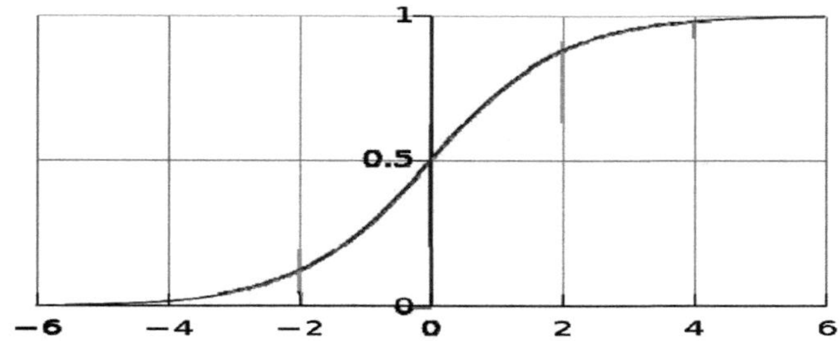

Figure 6.2 Curve shape of transfer function G2(s).

6.2.1 Description for Emotion Engine

Emotions in biological systems have been known to be caused by imbalances in biochemical neurotransmitters and (hormonal) secretions. Our intended model for emotion generation is based on the hormonal suppression/excitation dynamics in biological beings. It intends to emulate the similar dynamics. In each emotion segment, when the emotional intensity of function crosses a threshold it gets excited and transmits messages that suppress or stimulate other function activity. The signals can be either stimulate or suppress function onto whose territory they are sent. This behavior is embedded a priori in the system. Emotional expression is stimulated and suppressed from other function as follows [178]:

$$E^{s}_{\text{out}}(t+1) = E^{s}_{\text{old}}(t+1) + E^{s}_{\text{new}}(t) \tag{6.4a}$$

$$E^{s}_{\text{old/new}}(t) = \alpha \exp(-L) \tag{6.4b}$$

Individual stimulations are expressed exponentially, where L is the lifetime and α is a constant that could be selected to suit the application.

$$L = \omega\left\{\left(\frac{E^{s}_{\text{old}} - E^{s}_{\text{new}}}{R_m}\right)\right\} \text{ if } E^{s}_{\text{old}}(t) < E^{s}_{\text{new}}$$
$$= 0 \text{ (otherwise)} \tag{6.5}$$

where L is a positive constant and could be taken as unity.

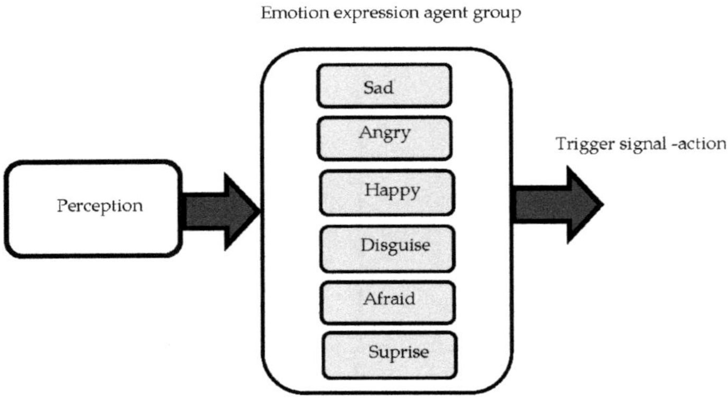

Figure 6.3 Architecture of each function in emotion expression.

6.2.2 Emotion Intelligent Controller Tuning by Artificial Immune Network Algorithm

The coding of an antibody in an immune network is very important because a well-designed antibody coding can increase the efficiency of the controller. As shown in Fig. 6.4, three type antibodies are discussed in this paper: (1) antibody type 1 is encoded to represent only α gain in the 3-DOF PID controller; (2) antibody type 2 is encoded to represent I gain; (3) antibody is encoded to represent β gains. The value of the k locus of antibody type 1 shows α gain allocated to route 1. That is, the value of the first locus of antibody type 1 means that α gain allocated to route 1 is obtained by 2 [64–66].

α	2	1	0.5	•••	0.2	0.1
β	2	1	0.5	•••	0.2	0.12
χ	2	1	0.5	•••	0.2	0.1

Figure 6.4 Allocation structure of α, β, χ gain in antibody.

On the other hand, the k locus of antibody 2 represents β gain for tuning of the 2-DOF PID controller. Here, the objective function can be written as follows:

$$\delta_i = \sum_{n=1}^{z}\{(L_n - L_n^{\text{object}})\}^2 + \zeta f_n$$

$$L_n = \sum_{i=1}^{P}(R_i I_{in})$$

$$f_n = \begin{cases} 0: L_n \le L_n^{\text{limit}} \\ 1: \text{Otherwise}, \end{cases} \tag{6.6}$$

where
δ_i is the objective function, z the number of processes for obtaining an optimal PID gain, L_n the optimal level in process for selection of an optimal gain, L_n^{object} the target optimal value in process for selection of an optimal gain, ζ the penalty constant, f_n the penalty function, P the number of route for selection of an optimal gain, R_i: gain level in route i, $I_{i,n}$ the subsidiary function, and L_n^{lim} the limit speed in PID gain.

The procedure for tuning is given in the following six steps:

Step 1: Initialization and recognition of antigen: The immune system recognizes the invasion of an antigen, which corresponds to input data or disturbances in the optimization problem.

Step 2: Product of antibody from memory cell: The immune system produces the antibodies that were effective in killing the antigen in the past. This is implemented by recalling a past successful solution from the memory cell.

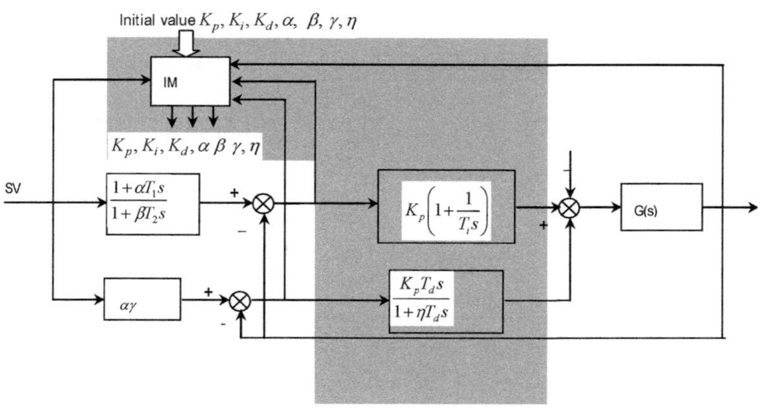

Figure 6.5 Immune-based tuning of the 3-DOF-EI-PID controller with emotion function.

Step 3: Calculation for searching an optimal solution.

Step 4: Differentiation of lymphocyte: The B-lymphocyte cell, the antibody that matched the antigen, is dispersed to the memory cells in order to respond to the next invasion quickly.

Step 5: Stimulation and suppression of antibody: The expected value η_k of the stimulation of the antibody is given by

$$\eta_k = \frac{m_{\phi k}}{\sigma_k}, \tag{6.7}$$

where σ_k is the concentration of the antibodies and is represented by

$$\sigma_k = \frac{\text{sum of antibodies with same affinity as } m_{\phi k}}{\text{sum of antibodies}} \tag{6.8}$$

Step 6: Stimulation of antibody: To capture the unknown antigen, new lymphocytes are produced in the bone marrow in place of the antibody eliminated in step 5.

6.2.3 Simulation and Discussion

Table 6.1 shows the comparison of tuning results in 3-DOF-EI-PID and shows that immune-based tuning is close to the best solution.

Table 6.1 Tuned gain by immune network

	Tuning methods			Response		
Item	K_p	T_i	T_d	t_r	t_s	**OV**
Z-N-PID	34.30	0.27	0.08	0.0774	1.2356	63.2
PI	23.23	0.38	—	0.1299	2.338	77.2
ISTE- ID	40.86	7.53	0.04	0.1441	4.7563	6.8
PI	11.90	8.044	—	0.1605	1.6713	18.6
IM	33.85	21.67	0.08	0.0263	2.4402	9.3

The mechanism of emotion function is biologically very complicated. That is, when the emotional intensity of a B-function crosses a threshold, it gets excited and transmits messages that suppress or stimulate other segment activity.

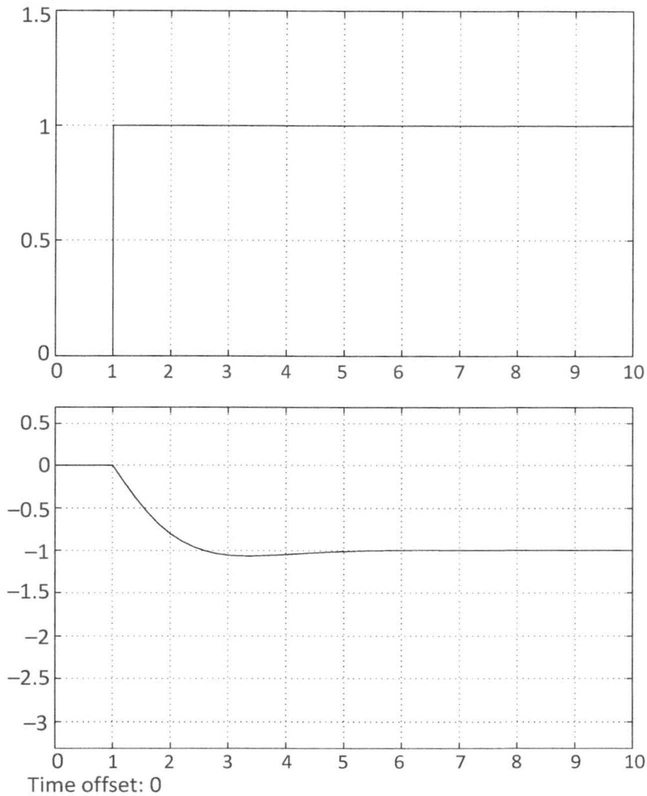

Figure 6.6 Simulated result with emotion function.

The messages on reaching other B-function territories stimulate or inhibit them, creating a positive or negative effect on the current emotion of the B-function. These messages form metaphors for biological neurotransmitters. They also have a life-time stamped on them by their creators at the time when they emanate, which is proportional to the change in the associated emotion intensity. The messages can either stimulate or suppress the B-function onto whose territory they are sent. The decision as to which B-function to stimulate and which to suppress has been made by intuitively classifying the affection of other emotions on a basic emotion, as positive or negative. It is also proportional to the input intensity to that B-agent among other factors discussed later.

6.2.4 Conclusions

The PID controller has been used in many areas. Also, the approach for achieving an optimal PID gain is broad and involves many tools such as Z-N, fuzzy, NN-fuzzy, and GA. The modified 2-DOF PID controller is designed by rearranging the 2-degree parameter to enable parameters of controller to fit into the industrial areas. The emotion function plays an important role in human being decision-making. This area represents an emerging technology. Here, this chapter suggests the approach for the application of emotion in industrial control and the other IT fields. For this purpose, we suggest an immune algorithm-based tuning method for the 3-DOF-EI-PID controller. Parameters *P*, *I*, and *D* encoded in the antibody are randomly allocated during selection processes to obtain an optimal gain for 3-DOF-EI-PID controller. The suggested controller can also be experimented effectively in the system.

6.3 Emotion Control of Induction Motor Using Disturbance Acceptance and Bacteria Foraging

6.3.1 Introduction

A proportional–integral–derivative (PID) controller has been widely used in the most industrial processes despite continual advances in the control theory. This is not only due to the simple structure which is theoretically easy to understand but also due to the fact that the tuning technique provides adequate performance in the vast majority of applications. However, it cannot effectively control such a complicated or fast-running system such as a disturbance control system, since the response of a plant depends on only the three parameters (*P*, *I*, and *D*) and its gain has to be manually tuned by trial and error in the industrial world. Most of the PID tuning rules developed in the past years use the conventional method such as frequency-response methods [30, 38–40] and disturbance rejection. Recently, the artificial intelligence approach is available for general linear control systems [6, 16–18] and bacterial foraging tuning [6].

On the other hand, the emotion function of the human mind can have an impact on decision-making, memory, action, and so on [26, 28, 29]. So, recently many articles and researchers have been interested in implementing such modeling, control, expression, and and so on [26–30]. Since natural selection of bacterial foraging tends to eliminate animals with poor foraging strategies for locating, handling, and ingesting food, optimization models can be provided for social foraging where groups of parameters communicate to cooperatively forage in engineering.

In the following section, we deal with emotion control by disturbance acceptance, and an intelligent tuning method of PID controller by bacterial foraging-based optimal algorithm is suggested.

6.3.2 Emotion PID Controller by Disturbance Acceptance

6.3.2.1 Condition for disturbance acceptance

In Fig. 6.7, the disturbance acceptance constraint can be given by [179]

$$\max_{d(t)\in D} \frac{\|Y\|}{\|d\|} = \left\| \frac{w(s)}{1+K(s,c)G(s)} \right\|_{\infty} \langle \delta. \tag{6.9}$$

Here, $\delta\langle 1$ is a constant defined by the desired acceptance level and $||\bullet||_{\infty}$ denotes the H_{∞}-norm, which is defined as

$$\|G(s)\|_{\infty} = \max_{w\in[0,\infty)} |G(j\omega)|. \tag{6.10}$$

The disturbance acceptance constraint becomes

$$\begin{aligned} \left\| \frac{w(s)}{1+K(s,c)G(s)} \right\|_{\infty} &= \max_{\omega\in[0\cdot\infty)} \left(\frac{w(j\omega)w(-j\omega)}{1+K(j\omega,c)G(j\omega,c)K(-j\omega,c)G(-j\omega,c)} \right)^{0.5} \\ &= \max_{\omega\in[0\cdot\infty)} (\sigma(\omega,c))^{0.5} \end{aligned} \tag{6.11}$$

The controller $K(s, c)$ is written as

$$K(s, c) = c_1 + \frac{c_2}{s} + c_3 s \tag{6.12}$$

The vector **c** of the controller parameter is given by

$$\mathbf{c} = \left[c_1, c_2, c_3\right]^T. \tag{6.13}$$

Hence, the condition for disturbance acceptance is given as

$$\max_{\omega \in [0 \cdot \infty)} (\sigma(\omega, c))^{0.5} \langle \delta. \tag{6.14}$$

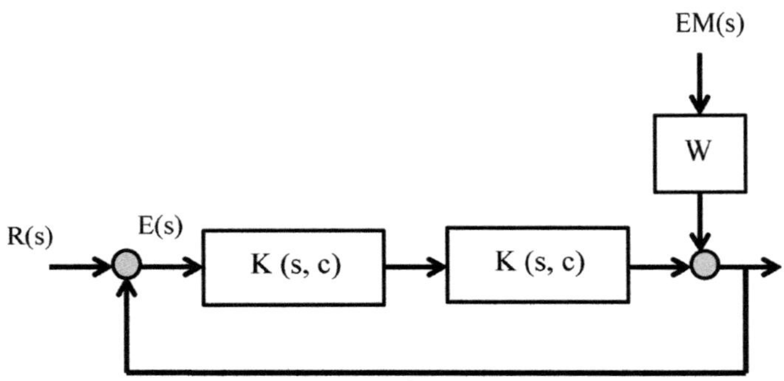

Figure 6.7 Emotion control system with disturbance acceptance.

6.3.2.2 Performance Index for disturbance acceptance emotion controller tuning

The performance index defined by ITWSE (integral of the time-weighted square of the error) is written by while to perform social foraging an animal needs communication capabilities,

$$E(s) = \frac{B(s)}{A(s)} = \frac{\sum_{j=0}^{m} b_j s^m}{\sum_{i=0}^{n} a_i s^n}. \tag{6.15}$$

Because $E(s)$ contains the parameters of the controller (c) and plant, the value of performance index, PI for a system of n-th order can be minimized by adjusting the vector **c** as follows [27]:

$$\min_{c} PI(c) \tag{6.16}$$

The optimal tuning proposed in this chapter is to find the vector **c**, such that the ITSE performance index, PI(c) is a minimum using bacterial algorithm and the constraint $\max_{\omega\in[0\cdot\infty)}(\sigma(\omega,c))^{0.5}\langle\delta$ is satisfied through real coded bacterial algorithms.

6.3.3 Optimal Computing of Bacteria Foraging

The characteristics of poor foraging can be applied to have an optimal solution through methods for locating, handling, and ingesting food. After many generations, a foraging animal takes actions to maximize the energy obtained per unit time spent foraging. That is, poor foraging strategies are either eliminated or shaped into good ones. So, this computing has capabilities of obtaining an optimal solution against environment. The foraging theory has been described in a number of works [115–117]. This section illustrates emotional function such as sad, anger, happy, or others by bacterial foraging. That is, to provide emotional function for user demand.

6.3.3.1 Emotion control by chemotactic behavior of *E. coli*

This section considers the foraging behavior of *E. coli*, which is a common type of bacteria as discussed in [4, 5]. Its behavior to move comes from a set of up to six rigid 100–200 rps spinning flagella, each driven as a biological motor. An *E. coli* bacterium alternates between running and tumbling. Running speed is 10–20 μm/sec, but it cannot swim straight. We can summarize the chemotactic actions of the bacteria as follows:

- If in neutral medium, alternate tumbles and runs, its action is having search.
- If swimming up a nutrient gradient (or out of noxious substances), swim longer (climb up nutrient gradient or down noxious gradient), its behavior seeks increasingly favorable environments.
- If swimming down a nutrient gradient (or up noxious substance gradient), then search action is avoiding unfavorable environments.

So, it can climb up nutrient hills and at the same time avoid noxious substances. The sensors it needs for optimal resolution are receptor proteins which are very sensitive and high gain. That is, a small change in the concentration of nutrients can cause a significant change in behavior. This is probably the best-understood sensory and decision-making system in biology.

Mutations in *E. coli* affect the reproductive efficiency at different temperatures and occur at a rate of about 10^{-7} per gene and per generation. *E. coli* occasionally engages in a conjugation that affects the characteristics of a population of bacteria. Since there are many types of taxes that are used by bacteria such as, aerotaxis (it are attracted to oxygen), light (phototaxis), temperature (thermotaxis), magnetotaxis (it can be affected by magnetic lines of flux. Some bacteria can change their shape and number of flagella, based on the medium to reconfigure in order to ensure efficient foraging in a variety of media. Bacteria can form intricate stable spatio-temporal patterns in certain semisolid nutrient substances. They can eat radially their way through a medium if placed together initially at its center. Moreover, under certain conditions, they will secrete cell-to-cell attractant signals so that they will group and protect each other. These bacteria can swarm.

6.3.3.2 Optimization function of bacterial swarm foraging

The main goal based on bacterial foraging is to find the minimum of $P(\phi)$ $\phi \in R^n$, not in the gradient $\nabla P(\phi)$. Here, when ϕ is the position of a bacterium, $J(\phi)$ is an attractant-repellant profile. That is, it means where nutrients and noxious substances are located, so $P < 0$, $P = 0$, $P > 0$ represent the status of nutrients, neutral medium, and presence of noxious substances, respectively [115–117]:

$$H(j,k,l) = \{\phi^i(j,k,l) \mid i = 1,2,\ldots,N\}, \tag{6.17}$$

This equation represents the positions of each member in the population of the N bacteria at the j-th chemotactic step, k-th reproduction step, and l-th elimination-dispersal event. Let $P(i, j, k, l)$ denote the cost at the location of the i-th bacterium $\phi^i(i, k, l) \in R^n$ [20, 21]. Let

$$\phi^i = (j+1,k,l) = \phi^i(j,k,l) + C((i)\varphi(j) \tag{6.18}$$

so that $C(i) > 0$ is the size of the step taken in the random direction specified by the tumble. If at $\phi^i(j + 1, k, l)$ the cost $J(i, j + 1, k, l)$ is better (lower) than at $\phi^i(j, k, l)$, then another chemotactic step of size $C(i)$ in this same direction will be taken and repeated up to a maximum number of steps N_S. N_S is the length of the lifetime of the bacteria measured by the number of chemotactic steps. Functions $P^i_c(\phi)$, $i = 1, 2, ..., S$, to model the cell-to-cell signaling via an attractant and a repellant is represented by [115–117]

$$P_c(\phi) = \sum_{i=1}^{N} P^i_{cc} = \sum_{i=1}^{N}\left[-L_{\text{attract}}\exp\left(-\delta_{\text{attract}}\sum_{j=1}^{n}(\phi_j - \phi^i_j)^2\right)\right] + \sum_{i=1}^{N}\left[-K_{\text{repellant}}\exp\left(-\delta_{\text{attract}}\sum_{j=1}^{n}(\phi_j - \phi^i_j)^2\right)\right], \tag{6.19}$$

where $\phi = [\phi_1, ..., \phi_p]^T$ is a point on the optimization domain, L_{attract} is the depth of the attractant released by the cell, and δ_{attract} is a measure of the width of the attractant signal. $K_{\text{repellant}} = L_{\text{attract}}$ is the height of the repellant effect magnitude), and δ_{attract} is a measure of the width of the repellant. The expression of $P_c(\phi)$ means that its value does not depend on the nutrient concentration at position ϕ. The model uses the function $P_{\text{ar}}(\phi)$ to represent the environment-dependent cell-to-cell signaling as

$$P_{\text{ar}}(\phi) = \exp(T - P(\phi))P_c(\phi), \tag{6.20}$$

where T is a tunable parameter. The model considers minimization of $P(i, j, k, l) + P_{\text{ar}}(\phi^i,(j, k, l))$, so that the cells will try to find nutrients, avoid noxious substances, and at the same time try to move toward other cells, but not too close to them. The function $P_{\text{ar}}(\phi^i,(j, k, l))$ implies that, with M being constant, the smaller the value $P(\phi)$, the larger the value $P_{\text{ar}}(\phi)$ and thus stronger the attraction, which is intuitively reasonable. In tuning the parameter M, it is normally found that when M is very large, $P_{\text{ar}}(\phi)$ is much larger than $J(\phi)$, and thus the profile of the search space is dominated by the chemical attractant secreted by *E. coli*. On the other hand, if T is very small, then $P_{ar}(\phi)$ is much smaller than $P(\phi)$, and it is the effect of the nutrients that dominates. In $P_{ar}(\phi)$, the scaling factor of $P_c(\phi)$ is given in an exponential form.

6.3.3.3 Computing procedure for optimization acceptance of emotion function

This section describes the method in the form of an algorithm to search optimal value of PID parameter [22–24].

Step 1: Initialize parameters n, N, N_c, N_S, N_{re}, N_{ed}, P_{ed}, $C(i)(i = 1, 2, ..., N)$, ϕ^i, and random values of PID parameter. Here,

n: Dimension of the search space (Each Parameter of PID controller),
N: The number of bacteria in the population,
N_c: chemotactic steps,
N_{re}: The number of reproduction steps,
N_{ed}: the number of elimination-dispersal events,
P_{ed}: elimination-dispersal with probability,
$C(i)$: the size of the step taken in the random direction specified by the tumble. The controller parameter is searched in the range of K_p = [0 30], T_i = [0 30], and T_d = [0 30].

Step 2: Elimination-dispersal loop: $l = l + 1$

Step 3: Reproduction loop: $k = k + 1$

Step 4: Chemotaxis loop: $j = j + 1$

Substep a: For i = 1, 2, ..., N, take a chemotactic step for bacterium i as follows.
Substep b: Compute ITSE (i, j, k, l).
Substep c: Let ITSE_{last} = ITSE (i, j, k, l) to save this value since we may find a better cost via a run.
Substep d: Tumble: generate a random vector $\Delta(i) \in R^n$ with each element $\Delta_m(i)$ $m = 1, 2, ..., p$, a random number on [−1, 1].
Substep e: Move: Let

$$\phi^i(j+1,k,l) = \phi^i(j,k,l) + C(i)\frac{\Delta(i)}{\sqrt{\Delta^T(i)\Delta(i)}}$$

This results in a step of size $C(i)$ in the direction of the tumble for bacterium i.
[substep f] Compute ITSE $(i, j + 1, k, l)$.
[substep g] Swim

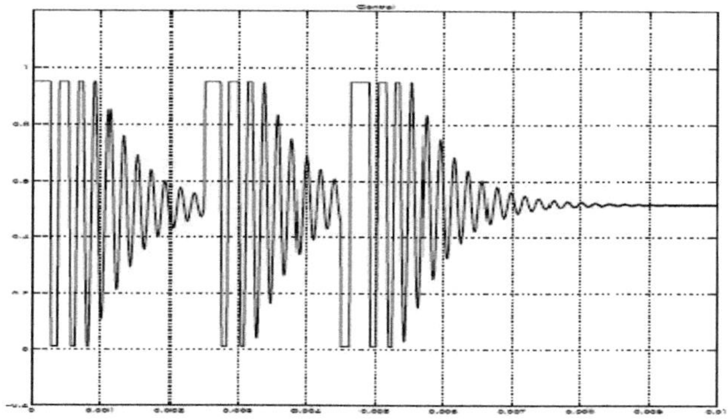

Figure 6.8 Traditional PID tuning with disturbance.

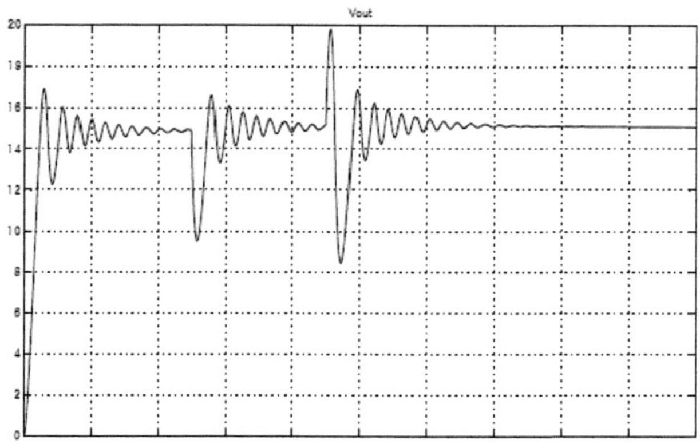

Figure 6.9 Bacterial foraging-based PID tuning with disturbance acceptance.

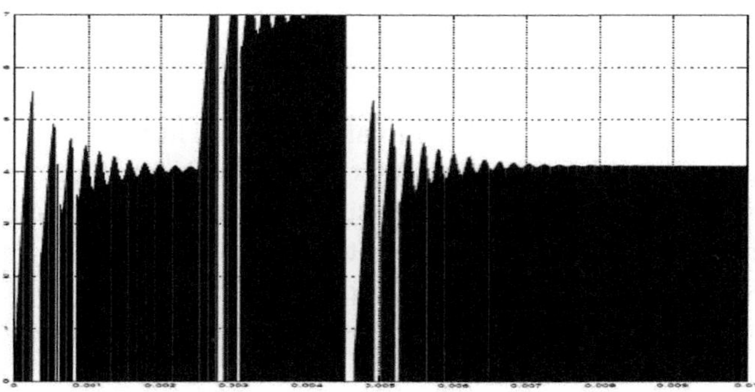

Figure 6.10 Source of emotion signal.

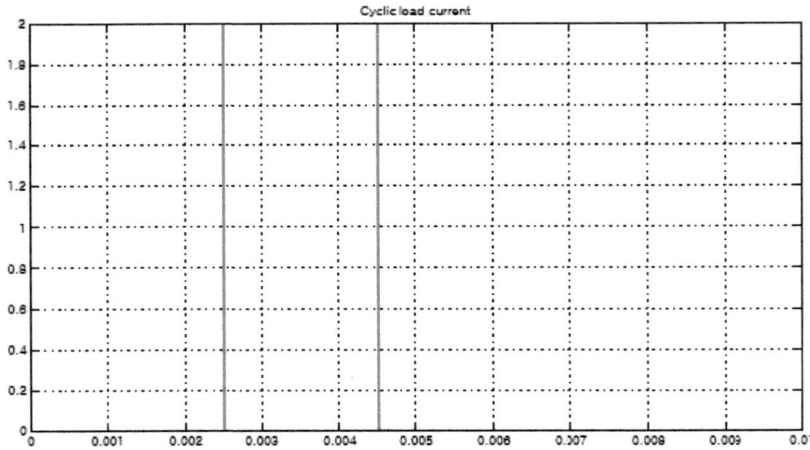

Figure 6.11 PID-based emotion signal.

6.3.4 Conclusion

In this section, we introduce the characteristics of bacterial foraging into emotion computing with disturbance rejection. Basically, many think that disturbance was not useful in the control system. However, in this section, we introduce that un-useful disturbance into our emotion control system positively by adding a traditional control structure. Of course, to use effectively, we need to do research in the real experimental and industrial systems in the future.

6.4 Emotion and Fuzzy Mood

6.4.1 Introduction

The emotion function of the human mind can influence decision-making, memory, and action. Also it plays an important role in artificial intelligence. Therefore, many articles and researchers have been interested in implementing such as modeling, control, expression, and robot. Also, an important thing of motivation for emotions research is based on a more autonomous and artificial robot. Basically, emotions as a fundamental part of the human being can have an effect on motivating actions and enriching experiences. It is now understood that a wide range of emotions play a critical role in every computer-related, goal-directed activity, from developing a 3-D CAD model and running calculations on a spreadsheet, to searching the Web. Many proposed a set of eight basic emotions, which are divided into four pairs of opposite states: joy–sadness; acceptance–disgust; fear–anger; surprise–anticipation. All these emotions are considered to be innate. This section suggests emotion function generation by a neural network to use intelligent control, action control, etc.

Many attempts at building models that emote have been attempted. We have tried to provide emotions on sensors to generate emotion by using fuzzy—positive ones suppressing the negative and vice versa and also stimulating their own kinds. Stimulations and suppressions have a magnitude, and lifetime can cause a temporal effect of the past in the present. Each fuzzy function forms a metaphor of an emotion-secreting unit or entity that has a restocking ability. We have used fuzzy logic to govern restocking of resource and the sampling of the environment. Sampling of the percepts from the sensors depends on a physiological effect. Stimulations, suppressions, emotion resource and a look-back before decay feature, all contribute to embed a deep and dynamic emotional nature into a machine. This section discusses the work done on the above and on the three emotions generated from percepts gathered by a robot moving along a path. This paper suggests emotion function generation by fuzzy agent to produce user requirement for emotion.

6.4.2 Emergent Emotion

Emotions can be of the positive or the negative type. Happiness, surprise, and enthusiasm are some of the positive ones, while anger, fear, and disgust constitute the latter. A neutral emotion is also defined which is neither positive nor negative but possibly signifies the calm state of a system. Positive and negative emotions can occur concurrently and cause one to yield to the other thereby changing the behavior of emoting. If we were to look at a shallow model, the predominant emotion will rule and be the cause that changes the behavior of the being. Such a system may be modeled by just finding the resultant of all emotions and then finding which emotion is predominant. Deep models, however, require extra dimensions. A deep model could be constructed based on which emotion has been predominant for a larger period of time in the past.

The first question to be addressed is how we decide the predominance. This is generally answered by referring to a quantification of emotion using the term emotional intensity. However, an apt and generalized definition of the intensity of emotion is still not around. A weighting, normalizing, and scaling mechanism could be used to derive an intensity of the emotion from sensors on a robot. Using these as emotion intensities, a resultant could be found and treated to be the winning emotion. These could in turn be used for emoting by the robot.

6.4.3 Physiologically Derived Emotion

Adrenalin, a hormone produced by the medulla of the adrenal glands, situated just above the kidneys, stimulates the heart and breathing rates and also increases the blood flow by dilating the blood vessels. It is known to be generated within the body during conditions of anguish, fear, and anger. We have attempted to model the input sensor sampling based on this phenomenon. The mood generated by the emotions of a system goes down making the system to start sampling at a faster rate as an attempt to regain over to a better mood. Low moods make the system desperate and force it to search for ways to make the mood better. The emotions churned out by the emotion engine decide the mood.

These emotions are processed by a fuzzy mood generation module to provide the current mood of the system. This fuzzy module takes in the three emotions, viz. happiness, fear, and anger, from the respective agents and resolves them to generate the internal resultant mood of the system.

6.4.4 Architecture for Emotion

Three fuzzy agents comprise the emotion engine. The inputs are derived from the robot's sensors, fuzzified and used to trigger the resource timer. They are also fed to the emotion engine via the latch which is enabled by the sampling timer which in turn is triggered by the resultant mood derived from the current emotions. Thus the rate of sampling changes inversely with the level of the mood. Figure 6.12 shows the architecture of the control system developed.

6.4.5 Operation for Emotion Producing

We have been able to make considerable progress in the development of a multi-agent emotion generating engine, as shown in Fig. 6.12.

Each agent generates one emotion concurrently with the others. A monitoring agent called the referee coordinates (dynamic controller). All emotions are grounded on the sensors mounted on the robot. To start with, we have used light, infra-red proximity, and sound sensors to provide inputs to the emotions engine. Emotion intensities are computed by each agent based on these inputs delivered to them by a referee agent and sent back. All agents compute the emotion intensities concurrently.

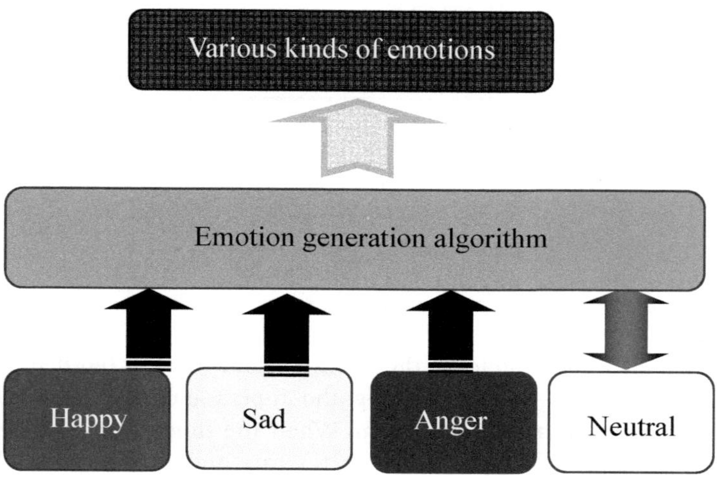

Figure 6.12 The architecture of fuzzy controller for emotion.

We have generated and sensed the mood of the environment in which the robot is situated [180]. Using this together with the rewards and penalties the robot receives from the environment, a fuzzy system generates the timing required to trigger the resource replenishing mechanisms within the agents.

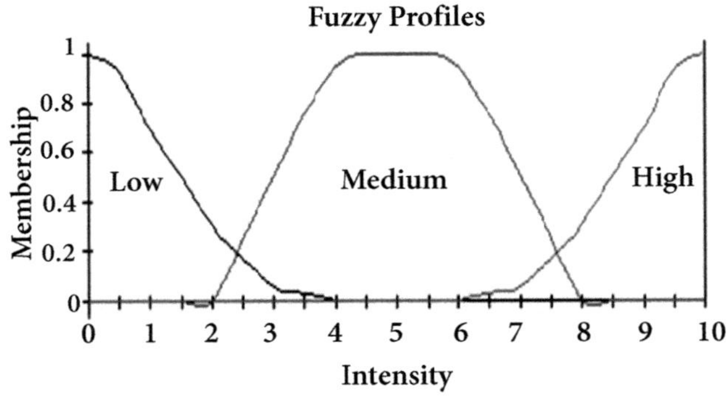

Figure 6.13 Fuzzy function for fuzzy mood.

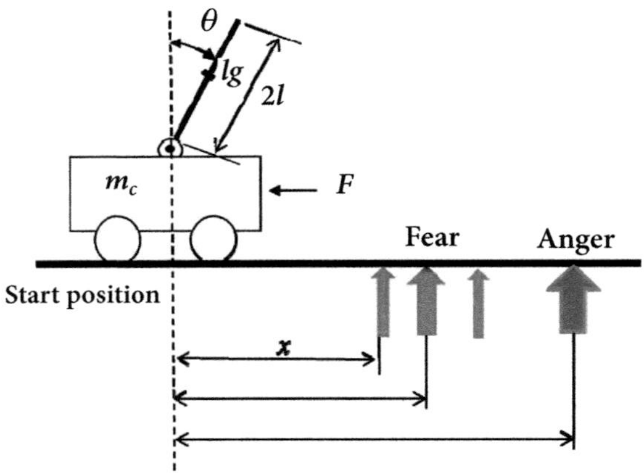

Figure 6.14 Experiment system for fuzzy mood in nonlinear system.

We also have modeled the agent system using a bio-inspired mechanism. The rate of sensing the inputs increase much the same way as in a human system. When the mood of the system is low, sensing takes place at a faster rate and vice versa in the other case. Mood is generated by a fuzzy logic–based module that fuzzifies the emotions generated within as shown in Fig. 6.13. The mood then decides the timing of a trigger that controls the sensing rate. Emotion decay has been modeled to be exponential with a temporal component to make the system exhibit a depth.

Stimulations and suppressions from positive and negative emotion generating agents tend to stimulate or suppress the emotion intensities. Both stimulations and suppressions have a magnitude and a life time. The former depends on the input, while the latter allows them to sustain their effect on the other agent for an extended period of time thereby adding to the depth of the emotional model. Tests were carried out on a partially simulated linear path. The results of the tests have been depicted. A schematic sketch of the robot moving along the intensity gradient is shown in Fig. 6.14.

6.4.6 Results

An internal mood for a robotic system using values sensed from its environment was quantified. Processed sensor values which act as metaphors for the external mood or ambience are responsible for the manner in which the robot emotes. In this section, we have grounded the external ambience sensing mechanism onto real sensors—one for each emotion. The inputs—happy, fear, and anger—could be individually generated based on a set of sensors with more sophisticated signal conditioning. A concept of emotional resource has been introduced and its significance cited using the relevant graphs. This resource is charged based on the external mood or the detected ambience and the rewards and penalties as the robot moves in an assigned path. This is very similar to what a human experiences.

A happy company always boosts positive emotions. We have used a fuzzy module to realize this. Emotions are generated by a multi-agent system. One agent dedicatedly generates an emotion. The multi-agent technique allows concurrent emotion generation.

This technique can provide the end users a facility to run the agents in those processors that are close to associated sensors. We may also run them in other systems that take part in emotion generation within a network. The structure of the system is also scalable. Figure 6.15 shows the variation of the various inputs, emotions generated, mood, rewards and penalties and the resource with time. As can be seen there is a temporal effect in the generated emotion as also certain depth and nondeterminism. The emotion agents make all the three emotions together and feed another fuzzy module. This module stimulates or suppresses sampling. The resultant mood of the system is delivered by this fuzzy module which controls the sampling of the exterior.

Mood is high when things are good and thus the sampling takes place at a slower rate. Thus the system is lazy. When the mood level falls, things are reversed. The resultant mood falls, and the system becomes more alert. Many agents can be attached to make a finer set of emotions. A robot can be made to assess and circumvent a problem using the mood effect. This can assist a robot to react accordingly to keep itself away from dangerous situations that cause low moods.

Table 6.2 Fuzzy associative memory for resource timer

Reward	Penalty	In-mood	EM T-value
Low	Low	Low	Moderate
Low	Low	Medium	Moderate
Low	Low	High	Low
Low	Medium	Low	Very (High)
Low	Medium	Medium	High
Low	Medium	High	Moderate
Low	High	Low	Very ($High^+$)
Low	High	Me	$High^+$
Low	High	High	High
Medium	Low	Low	Low
Medium	Low	Medium	Very (Low)
Medium	Low	High	Low
Medium	Medium	Low	Moderate
Medium	Medium	Medium	Moderate
Medium	Medium	High	Low
Medium	High	Low	$High^+$
Medium	High	Medium	Very (High)
Medium	High	High	High
High	Low	Low	Low^-
High	Low	Medium	Very (Low^-)
High	Low	High	Very (Low^-)
High	Medium	Low	Moderate

High	Medium	Medium	Moderate
High	Medium	High	Low
High	High	Low	High+
High	High	Medium	Very (high)
High	High	High	Moderate

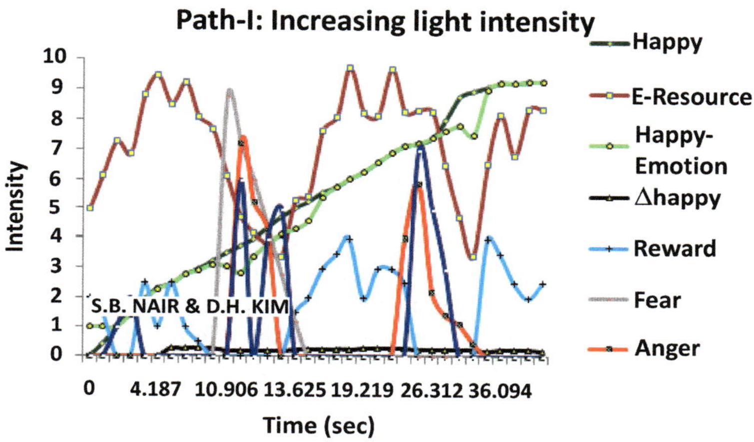

Figure 6.15 Experiment results.

6.4.7 Conclusion

The work in the area of robotics always requires that a real robot is there within the loop. It is useless to merely propound theories and not see them working. The lack of real robots has greatly retarded the progress of such research. Given real robots would have made a great deal of impact by fuzzy and mood. Further, more emotions will be added to test the scalability of the system. The sensors used to provide inputs and perceive the ambience of the environment too are currently limited. If more sensors can be mounted on the robot, a far better picture of the outside world could be gained by the emoting system. This will allow the generation of a richer and more realistic set of emotions. Bio-inspired techniques of emoting and learning behaviors could also be looked into. The robot can be situated in different environments and the corresponding emotional effects may express in different style.

6.5 Fuzzy Rule–Based Voice Emotion Control for User Demand Speech Generation of Emotion Robot

6.5.1 Introduction

Some psychologists suggested and studied emotion function as itself not in engineering. As voice emotion is one of them, it is an important communication in human-to-human interaction, teaching, and others. In these cases, emotional factors in the gender of the speaker, age, language, and stress can be influenced from the voice features of speaker [181, 182]. Usually, the influence of emotion in speech can be influenced by pitch frequency, power density, precise time alignment, and so on. Because voice parameters cause significant changes in the emotional states of voice, variation of emotion parameters is the most important for emotional generation for robot. Many studies illustrate a high correlation between some statistical data of speech and the emotional state of the speaker [181–183]. Among these research studies, the most popular are related to pitch, energy, articulation, and spectral shape. For example, sadness has been associated to low standard deviation of pitch and slow speaking rates, while anger usually implies higher values of pitch deviation and rate [183]. Most of the efforts done so far in emotion recognition are based on determining in voice recognition. That means how we have to generate emotional function for user demand. That is, we have to consider how we deal with these kinds of measure to get the information about the emotional state in robot or application. However, there are some reasons for the lack of interest in the vocal communication of emotion because it is not easy for psychologists to learn signal processing software and analysis of a mathematical training, and also we have to consider complicated noise in order to properly analyze speech.

This section suggests the way of fuzzy rule–based emotion function generation for user demand emotion function in robot.

6.5.2 Intelligent Control of Voice Emotion

In the field of human–robot interaction based on automatic speech recognition, recognition of emotion is challenging and important facts for robot's action. From the human perspective, a robot endowed with an emotion function should be capable of communicating an affective interaction with users. That is, it must have the ability to perceive, interpret, express, and regulate emotions. For these conditions, interacting with a robot would be more similar to communicating with humans and should be done more effectively. From the robot perspective, recognizing the user's voice emotional state is one of the main requirements for robot's moving to successfully interact with humans. Identification of expressiveness and voice emotion statuses should be improved for the understanding of the meaning conveyed by the communication process between robot and human being, as well as also provides a basic communication method with emotion and recover differentiating between satisfaction and dissatisfaction in the robot.

To generate emotion function for robot, we should express the characteristics of emotion state in voice. Intensity, energy, and pitch signal waveform of every sound has information about emotion because it depends on the tension and vibration of the vocal folds. The effect of basic emotions on the human speech is something that can alter the sound of the voice. The speech waveform can have an effect on variations voice patterns even slightly changes in physiological voice. So, we can create a number of methods to control voice by changes of pitch, frequency-domains in the time. Most of the methods have been useful for speech coders who have been trying to generate human speech for robot. In the time-domain parameters of the voice, some researchers can interpret the intensity and rate of speech as well as the level of the voice [185, 190]. In the frequency-domain, one can give the characteristics of the frequencies contained within the voice. Scherer [190] identifies the changes to the fundamental frequency of pitch in the voice, the intensity of the voice, and the energy distribution within the voice spectrum. Energy can affect the emotion pattern of the voice. So, the frequency, pitch, intensity, and speed of the voice are important factor for controlling emotion in robot. The important characteristics of the human voice signal should have been studied for user demand and its signal has been modified by certain emotions for robot. The main variables of the voice sound are the fundamental frequency range, the pitch variations, and the intensity. Usually, the F0 is the biggest effect on the listeners, with a narrow F0 suggesting an emotional state such as sadness [188–190].

In ref. [190], speaking rate has sad voice in range 150, angry in 179.

- F0: sad in smallest range, angry in the highest.
- Pitch level: happy in the highest, sad in the lowest.
- F0 slope: angry and happy in a high descendent.
- Spectral tilt: a lower tilt value increases the high frequency.

In ref. [14], a wide F0 is judged as expressing high arousal, producing attributions of strong negative emotions such as annoyance or anger. Similarly, high intensity in the voice signal was perceived negatively, associated with aggressive attitudes, whereas short-voiced segments, uttered with a fast speech rate, were interpreted as being joyous in their nature, as opposed to slow speech rate segments with a long duration, which were perceived to be akin to emotional sadness. When emotions are expressed through speech, high arousal emotions, such as hot anger, panic fear, despaired sadness, and elated joy, often are associated with a raised voice, fast speech rate, and higher pitch, when compared to low arousal emotions, such as cold anger, depressed sadness, calm joy/happiness and anxious fear. Also, pitch frequency is an important parameter in voice systems. Therefore we can produce user-demanded emotion in voice by controlling the effect of pitch frequency and its slope due to emotion can be explored for emotion control. That is, if we only use a few features such as energy, pitch, and intensity, we can produce or classify emotions.

6.5.3 Voice Emotion Intelligent Controller Tuning by Fuzzy Rule

6.5.3.1 Pitch contour

There are various kinds specifications in voice signal; pitch is one of them. The pitch has the key role for the perceptual sensation of how high or low certain tones are. Pitch function can be broken down into two areas: pitch period and pitch frequency. However, to express acoustic sound for situation, the estimation of the peak frequency in the spectrum should be introduced. These peaks, according to the evolution of their frequency and amplitude in speech time, should be considered according to the frequency relations among tones.

6.5.3.2 Acoustic energy

In the emotional state of anger or for stressed speech, Rainer and Klaus [191] defined that the fast air flow causes the false vocal folds providing additional excitation signals other than the pitch. In ref. [16], for the energy and pitch, the overall recognition rate decreased greatly, and the most emotions could be distinguished as only one or two emotions. The authors conclude that the energy and pitch features depend on the magnitude of the voice or the distance and are therefore not robust in a noisy environment. From this definition, we can control emotion situation by using intensity. The speech energy can affect their emotion status significantly.

6.5.3.3 Vocal tract features

The shape of the vocal tract is modified by the emotion status [191]. The formants, the cross-sectional areas, and frequency can be varied with situation for emotional status. However, we do not consider this term in this chapter.

6.5.4.4 Frequency

In ref. [190], the spectral center values are determined by the frequency of "*f*," not by the emotion. In other words, if someone speaks a sentence with an emotional state, then the spectral center of this speech is determined by how many "*f*" sounds are uttered, not by the emotion state of the speaker. Hence, even if the emotion recognition system is sentence independent, it is necessary to study the variation of the emotion features by phonemes, because most emotion features vary by phoneme.

6.5.4 Fuzzy Controller Voice Emotion

Stimulations and suppression function have a magnitude, and lifetime causes a temporal effect of the speech functions. We have used fuzzy logic to govern functions and the sampling of the environment.

Emotions could be produced by controller and three basic functions. Emotions generated by the emotions of a system can apply for action of a robot or another actuation. These emotions are processed by a fuzzy mood generation module to provide the current mood of the system. This fuzzy module takes in the three emotions: happiness, fear, and anger from the respective basic function and resolves them to generate the internal resultant mood of the system. Three functions comprise the emotion engine. The inputs are derived from the sensors in robot, fuzzified and used to trigger the controller to generate emotion for user demand.

6.5.5 Simulation and Discussion

Figure 6.16 is the schematic structure of fuzzy controller for user demand emotion and Fig. 6.17 is the structure of member function. Table 6.3 shows the fuzzy membership function.

Table 6.3 Fuzzy rule table for mood

Happiness	Fear	Anger	Mood
Low	Low	Low	Calm
Low	Low	Medium	Bad
Low	Low	High	Very (Bad)
Low	Medium	Low	Bad
Low	Medium	Medium	very (Bad)
Low	Medium	High	very (Bad)
Low	High	Low	Bad
Low	High	Medium	very (Bad)
Low	High	High	very (Bad)
Medium	Low	Low	Super
Medium	Low	Medium	Calm
Medium	Low	High	Bad
Medium	Medium	Low	Calm
Medium	Medium	Medium	Bad
Medium	Medium	High	very (Bad)
Medium	High	Low	Bad
Medium	High	Medium	very (Bad)

Happiness	Fear	Anger	Mood
Medium	High	High	very (Bad)
High	Low	Low	Low
High	Low	Medium	Good
High	Low	High	Bad
High	Medium	Low	Good
High	Medium	Medium	Bad
High	Medium	High	Very (Bad)
High	High	Low	Bad
High	High	Medium	very (Bad)

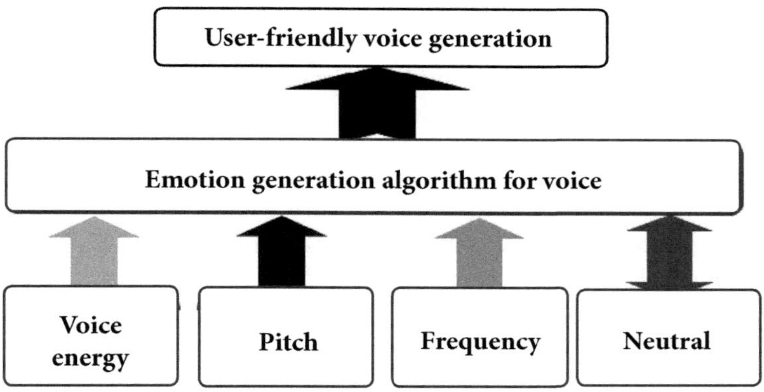

Figure 6.16 Fuzzy-based architecture developed for user demand emotion.

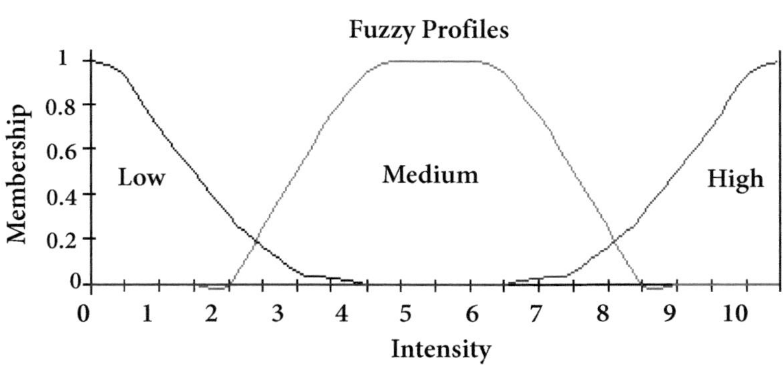

Figure 6.17 Fuzzy membership function for intensity.

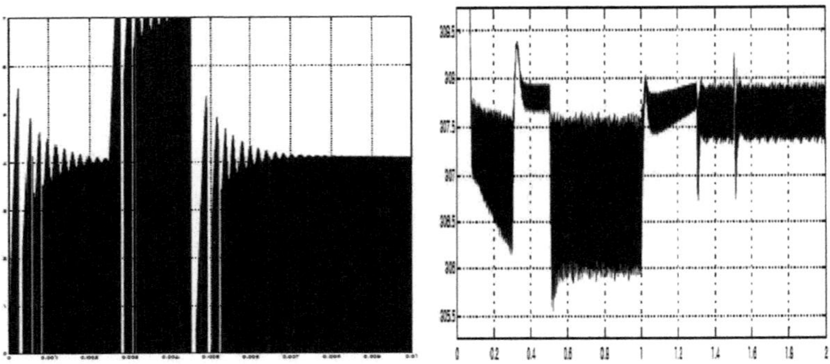

Figure 6.18 Source signal and signal pattern classified by control.

6.5.6 Conclusions

The voice emotion recognition is important for emotion robot and has a lot of useful applications in other areas. In real robot society, robots can be taught to interact with humans and recognize human emotions for communication. As the service robot is growing, robotic pets, delivery robots, and care robots, for example, should be able to understand the emotion situation and not only spoken commands.

That is, in an intelligent emotion robot system, detecting and adapting to user emotions should be considered. Without doing that, it is impossible to enable the systems not only to recognize the content encoded in user's response but also to extract information exchange about the emotional state of the user. Of course, we have to introduce to analyze spoken characteristics.

This chapter suggests the concept for producing emotional function by basic parameter and fuzzy control method. Of course, we have to study more experiments based on data and test by real voice signal.

Chapter 7

Hybrid System by AINFS and AINFNNS for Robust Control of Nonlinear System

7.1 Technology History and Artificial Intelligence

A significant challenge faced by engineers is the control of the dynamic characteristics in highly nonlinear industrial systems. It is well known that most industrial control systems exhibits nonlinearities from such processes as dynamics of fluids, electronic circuits, processes, or the mechanics of complex systems [191, 192, 193]. Common techniques used for working with nonlinear systems include the method of Lyapunov functions [227], artificial networks [194], adaptive methods [193], disturbance rejection [229], or disturbance observer [228].

The intelligent control technologies developed for industrial applications led to the emergence of novel technologies [197, 202, 203, 205, 217, 224]. This field of research is highly multi-disciplinary, including disciplines such as systems control, operations research, artificial intelligence, information and signal processing, computer software, or production background [206, 208].

Fuzzy logic was first intelligent systems technique, followed by neural, neuro-fuzzy, evolutionary systems, and their derivatives

Tuning Innovation with Biotechnology
Dong Hwa Kim

ISBN 978-981-4745-35-2 (Hardcover), 978-1-315-36458-2 (eBook)
www.panstanford.com

[195, 196, 198, 199]. These techniques allowed the creation of intelligent solutions for a wide variety of industrial applications.

More recently, authors suggested a model of fuzzy neuron in which linear synaptic connections are replaced with a nonlinearity characterized by a membership function and a fuzzy neural network model [194]. Since neural networks or fuzzy neural networks can conveniently describe a nonlinear relationship between multi-inputs and multi-output while having a short leaning time compared with a conventional neural network, it was expected that this approach would lead to the development of intelligent systems or soft computing approaches. For this purpose, there exist many different neural network models, such as Multi-Layer Perceptrons (MLP), Hopfield Networks, Kohonen Networks, Adaptive Resonance Theory (ART), etc. They are widely used in real-world applications in various fields such as signal processing, speech recognition, intelligent control, or robotics [4, 6, 14, 34].

Numerous fuzzy algorithms have been recently studied by a fuzzy logic system (FLS) or fusing with BPNN In most cases of fuzzy logic and fuzzy-neural networks, grade of membership and weighting function must be tuned by an approximation or experience-based tuning method. To address this issue, and attempt was made to identify efficient and robust genetic algorithms (GAs) for generating the fuzzy rules and weighting function. It was found that GAs can construct a set of fuzzy rules that optimize multiple criteria [199]. The rules searched by GAs were randomly scattered and had an optimal shape for the fuzzy membership function. Both neural networks and immunity-based systems are biologically inspired techniques that have the capability of identifying patterns of interest. They use learning, memory, and associative retrieval to solve recognition and classification tasks. However, the underlying recognition and learning mechanisms are very different. For example, the immune system provides diversification instead of converging to local or global optimization. Moreover, it possesses self-organizing memory and remembers its categorizations over long periods of time [211, 212, 214, 216].

This chapter proposes a hybrid learning system based on AINFS (Artificial Immune Network Fuzzy System) and AINFNNS (Artificial Immune Network-Fuzzy Neural Network System) for the optimal control of a nonlinear system. The concentration of antibody in the AINS is adjusted through the fuzzy membership function. The optimal shape of the fuzzy membership function is provided by the concentration of antibody, depending on error (deviation) between the set point and the nonlinear system output. It is shown that both are satisfactorily obtained for control purpose.

7.2 Computing Process for Information Getting of AINS

Biological information processing systems of **AINS** (Artificial Immune Network System) such as human beings or other natural systems have many interesting functions, expected to provide numerous feasible ideas to engineering fields, especially in intelligent control or robotics [196, 201, 217, 218, 224]. Biological information in living organisms can be mainly classified into the following four categories: nervous system, genetic system, endocrine system, and immune system [210, 211, 215]. Among these systems, the nervous system and genetic system have already been applied to engineering fields through concepts like neural network and genetic algorithms [197–199]. However, so far little attention was given to the application of such paradigms as immune algorithms in engineering, despite their important modeling characteristics.

The AINS implements a learning technique inspired by the human immune system, which is a remarkable natural defense mechanism that learns about foreign substances as shown in Fig. 7.1. However, the immune system has not attracted the same kind of interest from the computing field as the neural operation of the brain or the evolutionary forces used in learning classifier systems [214]. The immune system is a rich source of theories and can inspire computer-based solutions. Some computationally relevant characteristics of the immune system are listed below [30, 31]:

- The learning rule of the immune system is a distributed system with no central controller, since the immune system consists of an enormous number and diversity of cells throughout our bodies.

- The immune system has a naturally occurring event-response system which can quickly adapt to changing situations and shares the property with the central nervous system.
- The immune system possesses a self organizing and distributed memory. Therefore, it is thus adaptive to its external environment and allows a PDP (parallel distributed processing) network to complete patterns against the environmental situation.

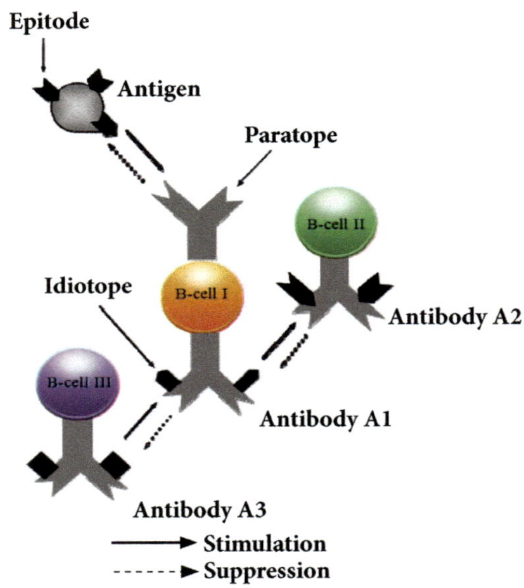

Figure 7.1 Information getting process in AINS.

In particular, the immune system has various interesting features such as immunological memory or immunological tolerance, since it can play an important role to maintain its own system in dynamically changing environments. The immune system is therefore expected to provide a new paradigm suitable for addressing dynamic problems involving changing environments, which usually expose the limitations of static systems.

Pioneering such approaches, Brooks has presented an architecture for behavior arbitration of autonomous robots [44]. He has argued that intelligence should emerge from mutual interactions among competence modules, and interactions between a robot and its environment.

Given that AINS can maintain its own system against dynamically changing environments, some researchers particularly focused on the new decentralized consensus-making system for engineering inspired by the biological immune system [201, 212, 214, 219].

There are several models to study both general and specific properties of the immunological phenomena. The main goal of these models is to deduce macroscopic principles of the system from the properties and interactions among the components of the immune system.

Moreover, the natural immune system is also a source of inspiration for developing intelligent methodologies toward problem solving, which allow the optimization of problems involving multiple criteria within dynamic environments.

A survey of the models that have been developed based on various computational aspects of the immune system and their applications to real world problems are available in [202], [220]. There are many intelligent approaches for their purpose as shown in Table 7.1.

Table 7.1 The characteristics of intelligence theory [29, 30, 33, 35]

Purpose of intelligence ◎: Best ●: Good								
Control	**Diagnosis**	**Pattern Recognition**	**Control**	**Plan**	**Decision**	**Design**	**Identify**	**Fore-cast**
Fuzzy	●	●	◎	◎	◎		◎	●
NN	●	◎	●				◎	◎
GA	●			◎	◎	◎		
Chaos	◎	●	●				◎	◎
Classical Control			●	●		◎		●

7.3 AINS-Based Anomaly Detection

Forrest et al. [26] developed a negative-selection algorithm for change detection based on the principles of self-nonself discrimination by T cells. In this approach, the *self* is defined as the normal pattern of activity of a system that we wish to protect or monitor. This algorithm exhibits many interesting properties in change detection. Dasgupta and Forrest [212] experimented with several data sets to investigate the performance of the negative selection algorithm for detecting anomaly in the data. Overall, the approach can be summarized as follows [219, 220, 221, 222, 225]:

- Collect times series (sensor) data, that sufficiently exhibits the normal behavior of a system (this may be raw data at given times, or average values over certain time intervals).
- Examine data to determine the range of variation in data patterns and choose the parameter according to the desired precision.
- Encode each value in binary (string), from within the observed data range.
- Select a suitable window size that can capture regularities in data patterns.
- Slide the window along the time rise and store the encoded string for each window as self, from which detectors will be generated.
- Generate a set of detectors that do not match any of the self strings.
- Once a unique set of detectors is generated from the normal data base of patterns, it can probabilistically detect any change (or abnormality) in patterns of unseen time series data.
- When monitoring the system, we are tracking parameters to encode new data patterns. If a detector is ever activated, a change in behavior patterns is known to have occurred and an alarm signal is generated regarding the abnormality. We use the same matching rule as was used in generating the original detectors.

This approach relies on a sufficient sample of normal data series that can capture the semantics of the data patterns and can represent the system behavior.

7.4 Dynamic Characteristic of AINS

7.4.1 The Response of AINS

The immune system has two types of response: primary and secondary. The primary response is the reaction when of the immune system when it first encounters the antigen. At this point, the immune system learns about the antigen, thus preparing the body for any further invasion from that antigen. This learning mechanism creates the immune system's memory. The secondary response occurs when the same antigen is encountered again. This response is characterized by a more rapid and more abundant production of antibody resulting from the priming of the B-cells (B-lymphocytes) as in the primary response. When a naive B-cell encounters an antigen molecule through its receptor, the cell gets activated and begins to divide rapidly; the progenity derived from these B-cells differentiate into memory B-cells and effector B-cells or plasma cells. The memory B-cells has a long life span and they continue to express membrane bound antibody with the same specificity as the origin parent B-cell [221, 222, 225].

7.4.2 Antibodies in AINS

The antibody molecule acts as a bridge between the cytotoxic cell and the target cell, subsequently causing the target cell to activate through a receptor. Antibodies are actually three-dimensional Y shaped molecules which consist of two types of protein chain: light and heavy. They also possess two paratopes which represent the pattern used to match the antigen. The regions on the molecules that the paratope can attach to are called epitopes. The molecules with antigenic peptide bound to them will be responsible for the interaction with T-cell receptors. The site that interacts with a T-cell receptor on an antigenic peptide is called epitode [225, 226].

7.4.3 Interaction between Antibodies

The antigen-antibody interaction is similar to that of enzyme-substrate interaction. The reaction between an antigen and antibody is of noncovalent type, where the antigenic determinants or epitodes interact with domains of the antibody molecule. The noncovalent interaction between antigen and antibody is brought about by hydrogen bonds, Van der Waals interactions, ionic bonds and hydrophobic interactions. Therefore, a strong affinity interaction should occur between antigen and antibody to form a stable complex. In Fig. 7.1, describing the interaction among antibodies is important to understand the dynamic characteristics of the immune system. Antigens stimulate the antibodies, consequently the concentration of antibody A1 and A2 increases. However, if there is no interaction between antibody A1 and antibody A2, these antibodies will have the same concentrations. Suppose that the idiotope of antibody A1 and the paratope of antibody A2 are the same. This means that antibody A2 is stimulated by antibody A1, and oppositely antibody A1 is suppressed by antibody A2 as Fig. 7.1. In this case, antibody A2 will exhibit a higher concentration than antibody A1. As a result, antibody A2 is more likely to be selected, as in this situation it has higher priority [211, 222].

7.4.4 Dynamics of AINS

In the immune system, the level to which a B cell is stimulated relates partly to how well its antibody binds to the antigen. We take into account both the strength of the match between the antibody and the antigen and the B cell object's affinity to other B cells, as well as its enmity. Therefore, the concentration of i-th antibody, which is denoted by δ_1, is calculated from the formula below, as suggested by [211, 214, 221]

$$\frac{dS_i(t)}{dt} = \begin{pmatrix} \alpha \sum_{j=1}^{N} m_{ji}\delta_j(t) \\ -\alpha \sum_{k=1}^{N} m_{ik}\delta_k(t) + \beta m_i - \gamma_i \end{pmatrix} \delta_i(t) \tag{7.1a}$$

$$\frac{d\delta_i(t)}{dt} = \frac{1}{1 + \exp\left(0.5 - \frac{dS_i(t)}{dt}\right)} \tag{7.1b}$$

where N is the number of antibodies, α and β are positive constants. m_{ji} denotes affinities between antibodies j and i (i.e., the degree of interaction), m_i represents affinities between the detected antigens and antibody i, respectively. Mutation γ is induced to prevent the occurrence of a local optimal in the computing process.

On the other hand, the information located in the lymphocyte population can be represented by [20, 21]

$$\Omega_j(N) = \sum_{i=1}^{S}(-x_{ij}\log x_{ij}) \tag{7.2}$$

where N is the size of the antibodies in a lymphocyte population, S is the variety of allele and x_j is the probability that allele i binds at locus j. Therefore, the mean information $\Omega_{ave}(N)$ for lymphocyte population is obtained from the following equation [34, 36]:

$$\begin{aligned}\Omega_{ave}(N) &= \frac{1}{M}\sum_{j=1}^{M}\Omega_j(N)\\ &= \frac{1}{M}\sum_{j=1}^{M}\left\{\sum_{i=1}^{S}-x_{ij}\log x_{ij}\right\},\end{aligned} \tag{7.3}$$

where M is the number of genes in an antibody.

The affinity $m_{\alpha\beta}$ between antibody α and antibody β is given by the following expressions:

$$m_{\alpha\beta} = \frac{1}{\{1+\Omega(\alpha\beta)\}} \tag{7.4}$$

$$\Omega(\alpha\beta) = H_s(x) = \left[f_1(x) + f_2(x) + f_3(x)\right],$$

where $\Omega(\alpha\beta)$ is the information obtained from the interaction of antibodies α and β. If $\Omega(\alpha\beta)$ = 0, then antibody α and antibody β match completely. Clearly the range of $m_{\alpha\beta}$ is between 0 and 1.

This chapter suggests a novel, immune system-based intelligent controller inspired by an immune system, using a membership function for the antibody concentration and an affinity function for the antigen.

7.5 Robust Control of Nonlinear System Using Hybrid System Composed of AINS and Fuzzy (AINFS)

7.5.1 Hybrid Structure of ANIS and Fuzzy System for Nonlinear System

7.5.1.1 Membership function

In this chapter, a fuzzy membership function is used to compute the concentration of an antibody in AINFS, depending on the characteristics of a nonlinear system as shown in Fig. 7.2.

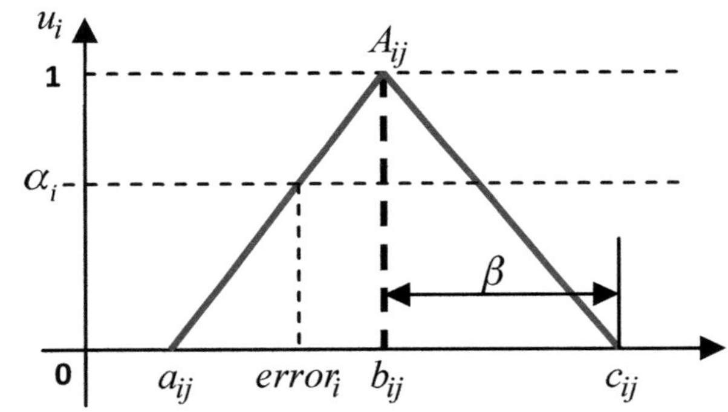

Figure 7.2 Relation of AINFS and fuzzy membership function

The triangle type membership described in Fig. 7.2 and Eq. (7.5) is used in this chapter.

$$A_{ij}(x_j) = \begin{cases} 1 + \dfrac{x_j - b_{ij}}{b_{ij} - a_{ij}} & : a_{ij} \le x_j < b_{ij} \\ 1 + \dfrac{b_{ij} - x_j}{c_{ij} - b_{ij}} & : b_{ij} \le x_j \le c_{ij} \\ 0 & : \text{otherwise} \end{cases} \tag{7.5}$$

The weight values w_i and controller output u_i are linked by Eq. (7.6). Deviation between reference and output is defined as an antigen from outside. Therefore, the purpose of the control is to ensure that this deviation should be zero by the weights w_i of the membership function.

$$u_i = \max(\mu_{A_{ij}}) \cdot w_{ij}. \tag{7.6}$$

Here, w_{ij} is weight function that antibody A_{ij} can have depending on nonlinear characteristics.

7.5.1.2 Affinity calculation

The concentration constant α of AINFS increases when the number of antigens in the outside increases.

Figure 7.3 shows the principle that the concentration $\alpha \geq \max(\mu_{A_{ij}})$ of antibody is generated through crossover with neighbor against the number of antigen, being computed by the error or deviation between the set value and output in nonlinear system. Then, the number β of antibodies is limited by fuzzy membership function.

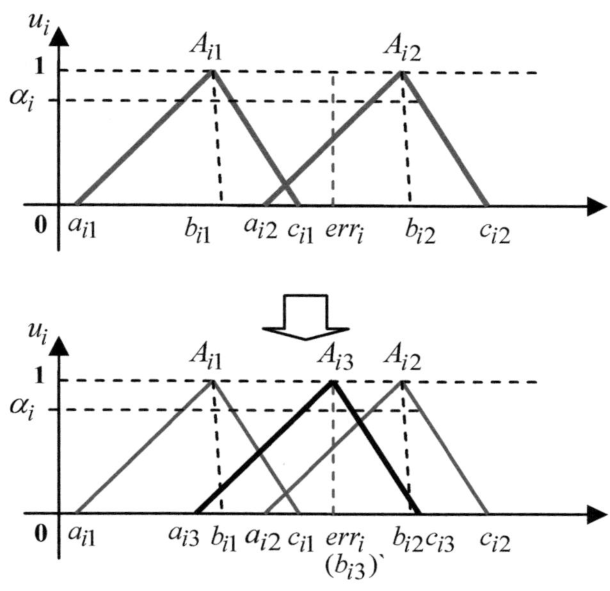

Figure 7.3 Generation of new antibodies.

Figure 7.4 illustrates the variation of the fuzzy membership function with the variation of antibodies β.

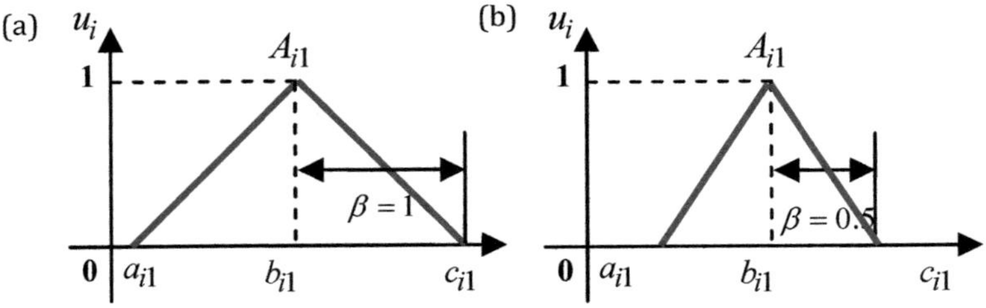

Figure 7.4 Variation of antibody with β.

7.5.1.3 Plant model and initial condition

The dynamic equation of the level control system is given by

$$\dot{h} = (V_{\text{in}} - V_{\text{out}}) / \text{Area},$$
$$\text{Area} = \pi * R^2 = A_k$$
$$V_{\text{out}} = K\sqrt{h},$$
$$V_{\text{in}} = f(u)$$
$$\dot{h} = \frac{f(u) - K\sqrt{h}}{A_k} \rightarrow \dot{h}A_k + K\sqrt{h} = f(u) \tag{7.7}$$

Where h is the level of tank, u the position of valve, Area refers to the area of tank, and K is the resistance of the pipe.

Figure 7.5 shows the shape of membership function on the interval [−40, +40] for initial membership against error in a immune network, and initial antibody number 20.

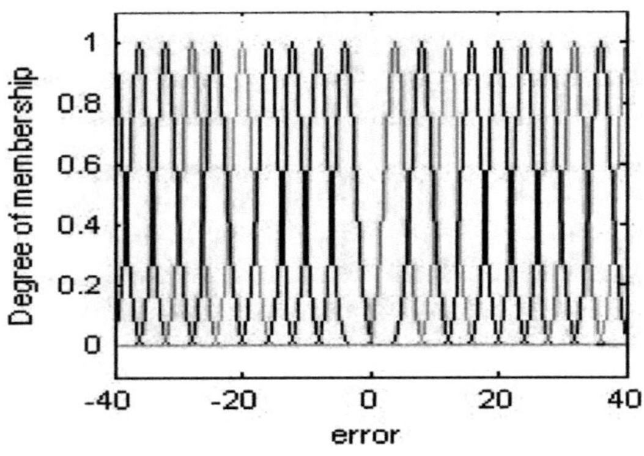

Figure 7.5 The shape of fuzzy membership by initial immune network (initial antibody no.: 20, gaussmf).

Figure 7.6 shows the response when the level of water varies from 0 to 5. The response is quickly stabilized at the required level, suggesting that antibodies have the ability to capture antigens and rapidly get rid of the error. In this example the number of initial antibody is 20, the immune system mutation constant γ is 0.1, the number of antibodies α is 0.5, and positive constant β is 1.5.

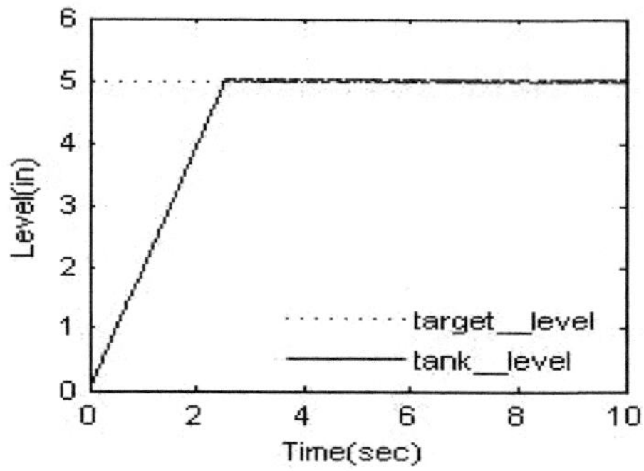

Figure 7.6 The output characteristics of controller (initial antibody no.: 20, $\gamma = 0.1, \alpha = 0.5, \beta = 1.5$, gaussmf).

Figure 7.7 shows the effects of increasing the number of antibody with time. In this case, antibodies are shown to have good information processing in time and efficiently remove errors for increased number of antibody. Eventually, there is no error in the control system after learning. The parameters here are: initial number of antibodies in immune network = 20, mutation of immune system γ = 0.1, the number of antibodies α = 0.5, positive constants β = 1.5, Gaussian membership function.

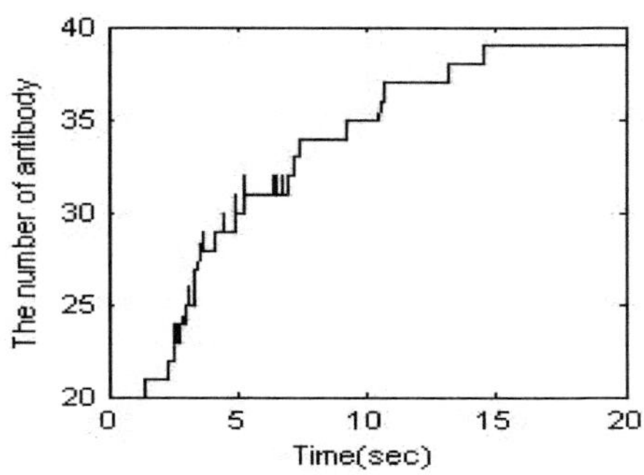

Figure 7.7 Variation of the number of antibody to time (initial antibody no.: 20, $\gamma = 0.1, \alpha = 0.5, \beta = 1.5$, gaussmf).

Figure 7.8 illustrates the shape of membership function after learning by AINS. It is the shape of membership function optimized from initial shape of Fig. 7.2.

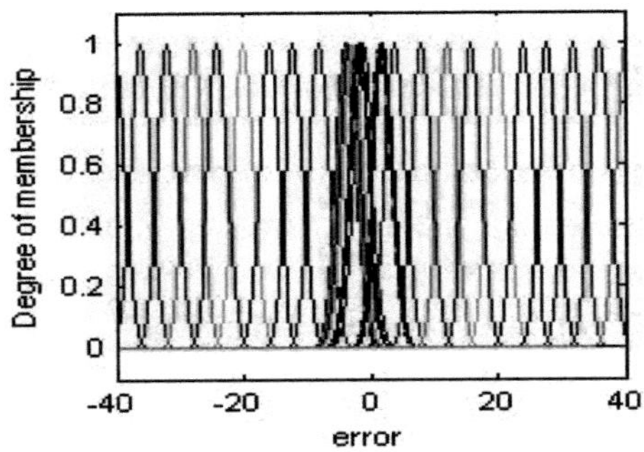

Figure 7.8 The shape of membership function to error by variation of antibody (initial antibody no.: 20, $\gamma = 0.1$, $\alpha = 0.5$, $\beta = 1.5$, gaussmf, end time = 20 sec).

7.5.1.4 Characteristics of antibody β

Figure 7.10 is the number of antibody to the variation of β (initial antibody no.: 20, $\gamma = 0.1$, $\alpha = 0.2$, gaussmf). For increasing values of β in Fig. 7.10, the convergence is faster until it saturates at $\beta = 50$. It means that when an antigen is captured by the antibody, a reaction is no longer possible, in accordance with the principle of AINS. Figure 7.10 shows the degree of membership to the variation of β (initial antibody no.: 20, $\gamma = 0.1$, $\alpha = 0.2$, gaussmf). The response level of the system depending on variation of β in AINS is depicted in Fig. 7.9.

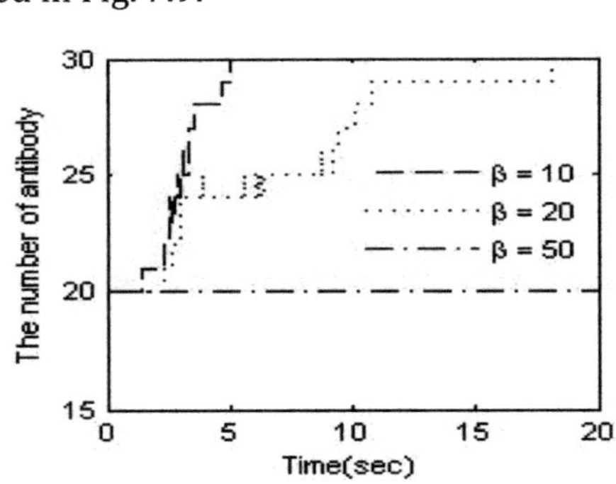

Figure 7.9 The number of antibody to the variation of β (initial antibody no.: 20, $\gamma = 0.1$, $\alpha = 0.2$, gaussmf).

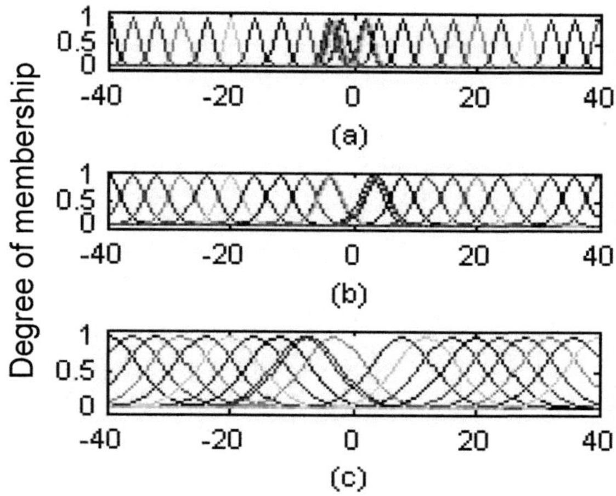

Figure 7.10 The degree of membership to the variation of β (initial antibody no.: 20, γ = 0.1, α = 0.2, gaussmf).

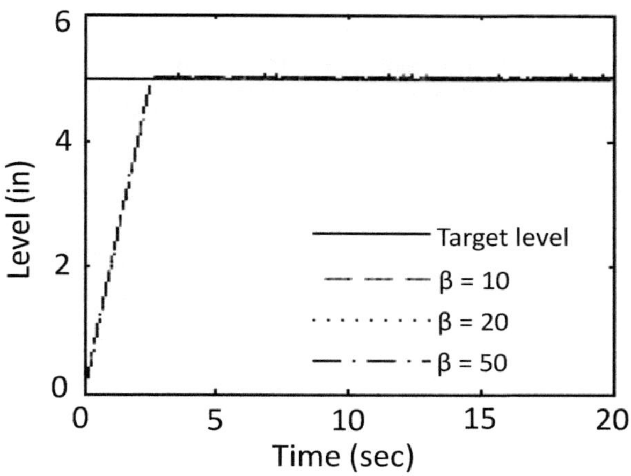

Figure 7.11 The output characteristic of controller with β (initial antibody no.: 20, γ = 0.1, α = 0.2, gaussmf).

7.5.1.5 Characteristics of concentration constant of antibody α

Figure 7.12 represents the degree of membership to error and α (initial antibody no.: 20, γ = 0.1, β = 1.5, gaussmf). When the concentration constant of antibody α is bigger (0.7), the shape of membership function can be tuned more quickly.

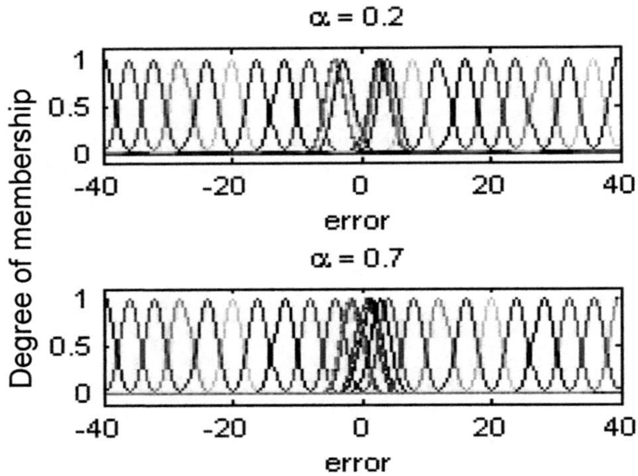

Figure 7.12 The degree of membership to error and α (initial antibody no.: 20, γ = 0.1, β = 1.5, gaussmf).

Figures 7.13 and illustrate the level characteristics to the variation of α to time, respectively. Clearly, when α is bigger, the level response converges faster to the target level the faster (initial antibody no.: 20, γ = 0.1, β = 1.5, gaussmf), as shown in Fig. 7.13.

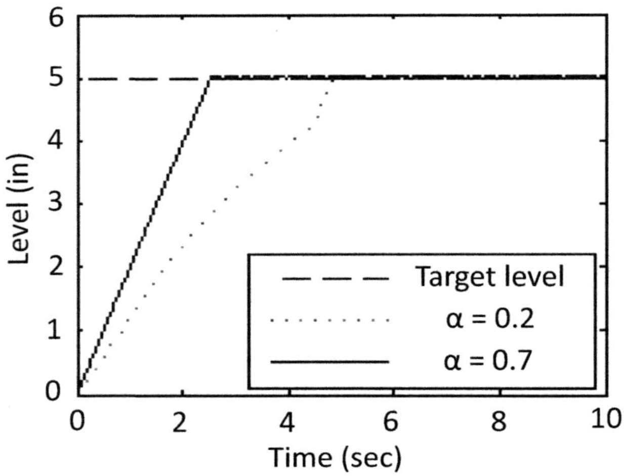

Figure 7.13 The level characteristic to the variation of α (initial antibody no.: 20, γ = 0.1, β = 1.5, gaussmf).

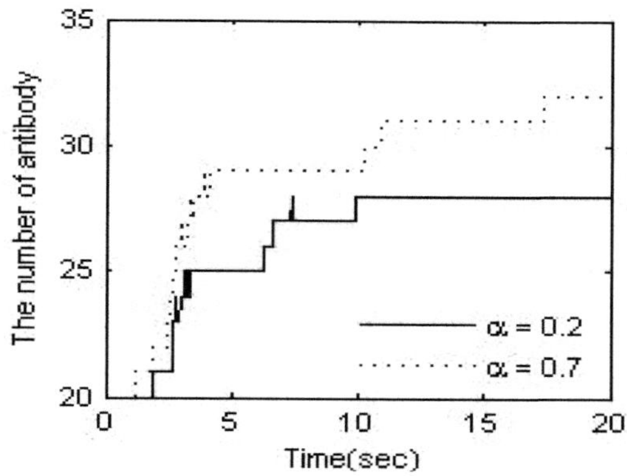

Figure 7.14 The variation of antibody No. α to time (initial antibody no.: $20, \gamma = 0.1, \beta = 1.5$, gaussmf).

As we can see, by changing the degree and the variation of γ, the antibody may improve the capability to quickly respond to antigen (error).

7.5.1.6 Characteristics of membership function shape to variation of the number of antibodies

Figure 7.15 is the variation shape of membership function to the variation of the number of antibody (a = 10, b = 30, c = 50, γ = 0.1, α = 0.5, β = 1.5, gaussmf). Figure 7.16 shows the characteristic to the level variation to the antibody number (γ = 0.1, α = 0.5, β = 1.5, gaussmf).

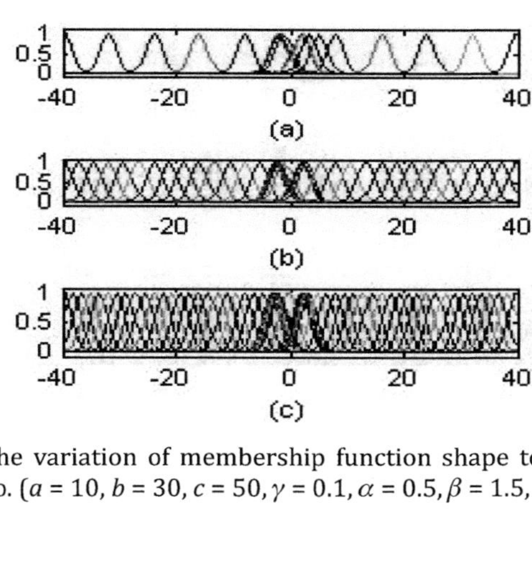

Figure 7.15 The variation of membership function shape to the variation of antibody no. ($a = 10, b = 30, c = 50, \gamma = 0.1, \alpha = 0.5, \beta = 1.5$, gaussmf).

The best response is when the number of antibody is 10. When the number of antibodies is over this mark, an over decay is happening, as illustrated in Fig. 7.16. Here the variation characteristics of the number of antibody to time is presented ($\gamma = 0.1, \alpha = 0.5, \beta = 1.5$, gaussmf).

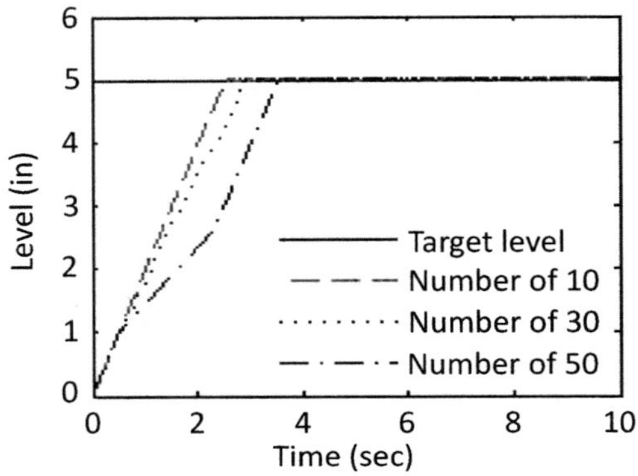

Figure 7.16 The characteristic to the level variation to the antibody No. ($\gamma = 0.1, \alpha = 0.5, \beta = 1.5$, gaussmf).

Figures 7.16 and 7.17 show the importance of the initial number of antibody choice for getting a good response. This is because some time is needed for the reaction of antibodies, especially in the case when the antibody count is small to the target control. Of course, it means that there is a possibility for improved performance, as antibodies have a good chance to react with antigen.

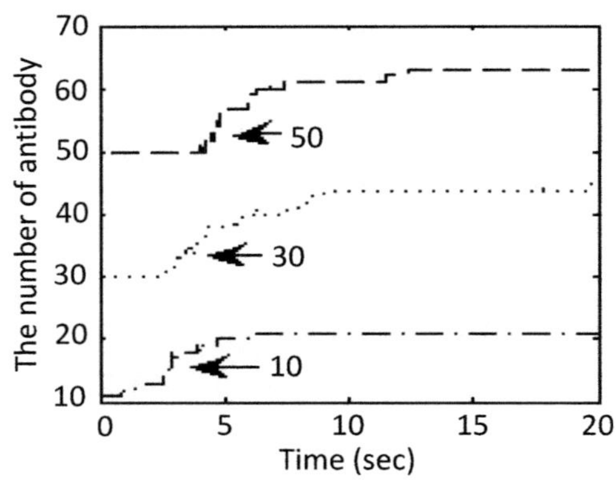

Figure 7.17 The number of antibody to time ($\gamma = 0.1$, $\alpha = 0.5$, $\beta = 1.5$, gaussmf).

7.5.1.7 Comparison of the output of the controller with others

In this chapter, other methods such as fuzzy and neural network are shown, to test the performance of control system by the concentration of antibody of Fig. 7.18. The AINS algorithm suggested in this chapter is found to compare well against other approaches.

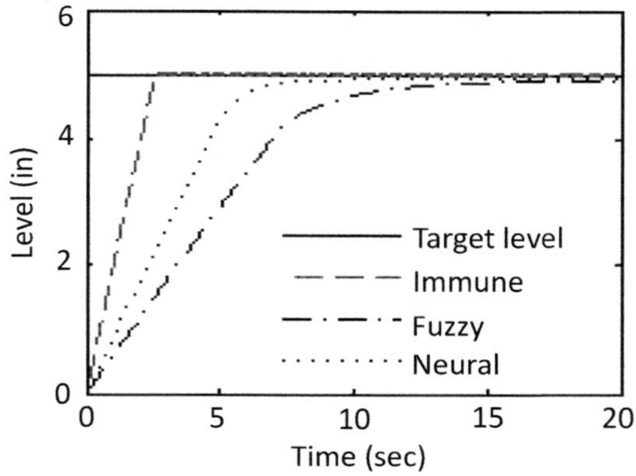

Figure 7.18 Comparison with fuzzy and neural network (initial antibody No.: 20, $a = 0.5, \gamma = 0.1, \alpha = 0.5, \beta = 1.5$, gaussmf).

7.5.2 Robust Control of Nonlinear Car System with Pendulum by AINFS

7.5.2.1 Fuzzy-based state space for pendulum system

The inverted pendulum has been used for testing control algorithms for nonlinear systems. This section provides a hybrid system composed of AINS and Fuzzy for optimal control of a car with pendulum system. For optimal control, the distance between the target position b and the initial position a is controlled by the resolution, α. The formula of the resolution is given by

$$\left|\min(a,b)+\left(\frac{\max(a,b)-\min(a,b)}{\alpha}\right)\times(i-1)\right|<|d_i|$$

$$|d_i| \leq \left|\min(a,b)+\left(\frac{\max(a,b)-\min(a,b)}{\alpha}\right)\times i\right|,$$

$$i=\alpha,\alpha-1,\alpha-2,\ldots,1. \tag{7.8}$$

Here d_i is the present position of the car. In our approach, we introduce the fuzzy method to divide the control partition of the state space of nonlinear systems depending on control characteristics. As Eq. (7.8) shows that the error is controlled by the number of the antibodies α, it is important to define the characteristics of the nonlinear control system shown in Fig. 7.19.

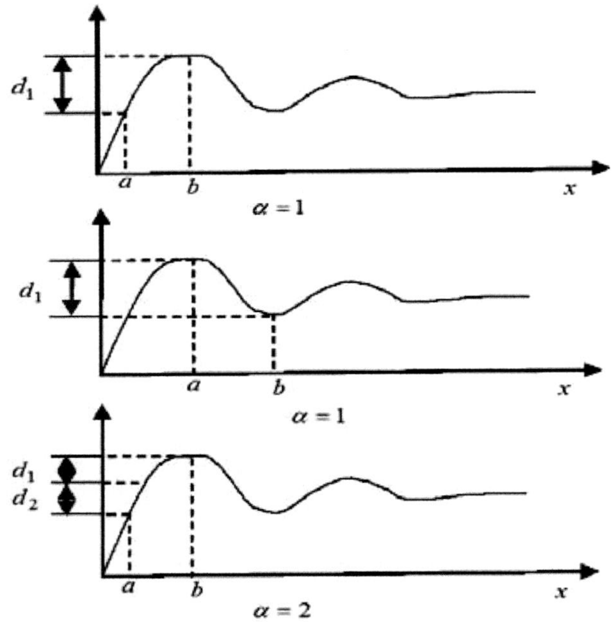

Figure 7.19 Variation of resolution depending on α.

We define the control error as

$$e_i = \int_{t=1}^{n} t(xc^2 + \theta^2)dt, i = \alpha, \alpha - 1, ..., 1. \quad (7.9)$$

Here, n is the number of the partition given by time before the system change, and the fuzzy output can be written as

$$f'_{i.i} = w_{i.i} \times f_i, i = \alpha,\ \alpha - 1, ..., 1. j = \text{a number of fuzzy rule}, \quad (7.10)$$

where $w_{i,j}$ is the gain to the fuzzy output f_j (j = *a number of fuzzy rule*) in the distance i(i = α, α−1, α−2, ..., 1). For example, if the number of fuzzy rules is 16, then (j = 1, 2, 3, ..., 6). The structure of antibodies for the nonlinear system is shown in Fig. 7.20. The affinity between the antibody and antigen is defined by $e_{i\,=\,1,\,2,\,3,\,\ldots,\,\alpha}$. The affinity between the antibodies is decided by the control error e_i.

$$\eta_i = \frac{\alpha}{\alpha + f(x)}, \tag{7.11}$$

$$f(x) = \sum_{i=1}^{a} x_i,\ x_i = \begin{cases} -1, \text{ if } e_i < e_{i+1}, \\ 0, \text{ if } e_i = e_{i+1}, \\ 1, \text{ if } e_i > e_{i+1}. \end{cases} \tag{7.12}$$

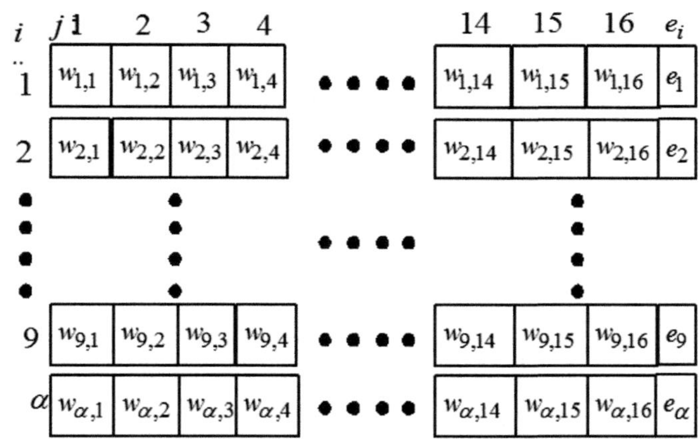

Figure 7.20 The structure of antibody for the nonlinear pendulum.

The antibody concentration for the optimal solution is

$$\Phi = \sum_{i=1}^{a} \eta_i \Big/ \alpha \tag{7.13}$$

We emphasize that the mutation and crossover operators used are introduced by the genetic algorithms.

7.5.2.2 Nonlinear characteristic of pendulum

Controlling the car's position is rather difficult, since the pendulum system has a very strong nonlinearity, as depicted in Fig. 7.21. In this chapter, the Sugeno fuzzy control system in Fig. 7.22 is employed. The angular velocity of the pendulum is denoted by $\dot{\theta}$, the speed of the car by $\dot{x}$, the gravity by $\dot{g}$, the external force τ, the general position is q, the general speed is $\dot{q}$, while $\ddot{x}$ is

$$\ddot{x} = \frac{F + m_p l\left[\dot{\theta}^2 \sin\theta - \ddot{\theta}\cos\theta\right]}{m_c + m_p}. \tag{7.14}$$

Also, θ is

$$\ddot{\theta} = \frac{g\sin\theta + \cos\theta\left[\dfrac{-F - m_p l\dot{\theta}^2 \sin\dot{\theta}}{m_c + m_p}\right]}{l\left[\dfrac{4}{3} - \dfrac{m_p \cos^2\theta}{m_c + m_p}\right]}. \tag{7.15}$$

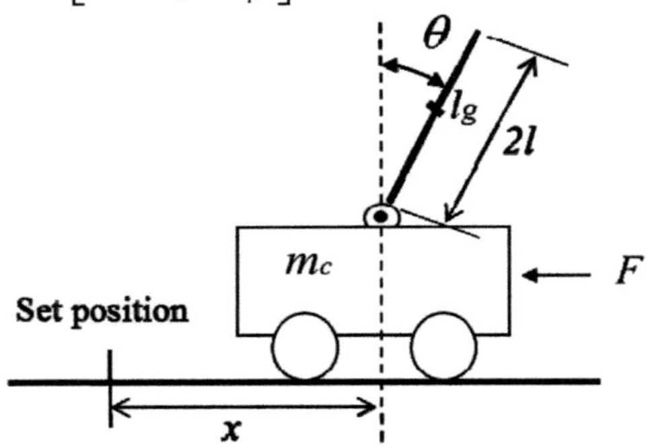

Figure 7.21 Structure of car with pendulum.

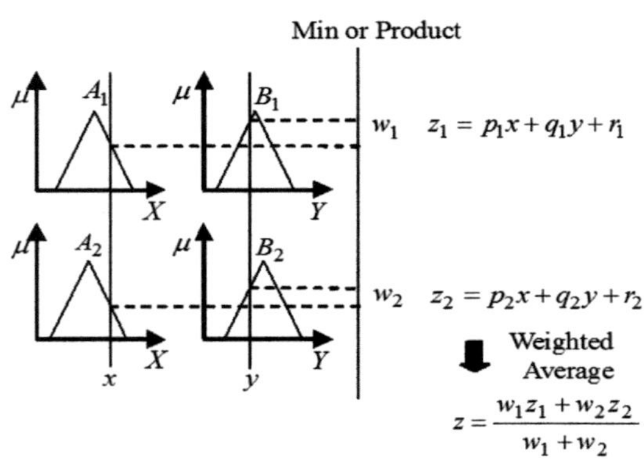

Figure 7.22 Sugeno fuzzy control model.

Table 7.2 Computed weight function

	w_1	w_2	w_3	w_4	w_5	w_6	w_7	w_8	e_i
1	1.7092	1.1366	1.8172	1.8651	0.9971	0.3958	0.4252	0.1375	2747.3
2	1.6256	1.3709	1.864	1.8614	1.1668	1.67	1.7432	0.2961	22.206
3	0.2030	0.6786	0.0061	0.7707	0.0365	0.3096	1.9885	0.4221	4.2395
4	0.9361	0.4963	1.2069	0.4873	0.7342	1.9955	1.7916	0.7562	0.6114
5	1.2857	1.2261	1.9534	1.9603	0.9868	1.7432	1.8806	0.5040	78.815

Φ: 5.0299

7.5.2.3 Controller design for car with pendulum

In this chapter, we simulate the pendulum car with the following specifications: the weight of the cart is m_c = 2 kg, while the length and mass of the pendulum are l = 0.5, m_p = 0.1 kg, respectively. Figure 7.23 shows the MATLAB/SIMULINK-based [47] simulation model for our fuzzy-immune controller.

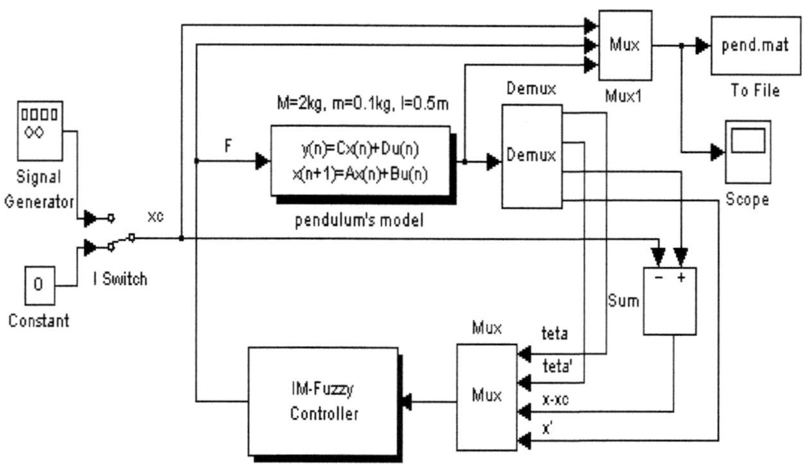

Figure 7.23 MATLAB/SIMULINK-based fuzzy-immune block for simulation.

As initial conditions we use θ = 0.1 *rd*, x = 0.5 *m*, with the reference position x = 0. The ranges of these variables are defined by the values $-0.3 < \theta < 0.3, -1 < \dot{\theta} < 1, -3 < x < 3, -6 < \dot{x} < 6$.

The membership functions for each variable are illustrated in Fig. 7.24.

Table 7.3 Membership shape depending on generation of AINFS

Item	X_1_min	X_1_max	X_2_min	X_2_max	δ	α	PI	E_PI
Ai	[0.46, 0.48]	[0.77, 0.81]	[45.0, 47.0]	[61.0, 63.0]	[0.001, 0.001]	[0.00001, 0.0004]		
A_0	0.4799677467	0.7789457406	46.247500655	62.2563049853	0.00100149345	0.0000354499	0.040311	0.27306
A_1	0.4799938583	46.2618954295	0.7814936556	62.4140905514	0.00100555325	0.0000465945	0.040598	0.27277
A_2	0.4799995994	46.2495033736	0.7825169491	62.5021071454	0.00100012016	0.0000197420	0.040491	0.27288
A_3	0.4799996948	46.2499020098	0.7813410294	62.4214891638	0.00100003433	0.0000420978	0.040452	0.27292
A_4	0.4600006103	45.0051097918	0.7700020980	61.0629330281	0.00268751019	0.0000279584	0.03265	0.28551
A_5	0.4600009918	45.0002574923	0.7700003814	61.2041646997	0.00262390100	0.0002727694	0.035923	0.28475
A_6	0.4600003242	45.0000438690	0.7700003814	61.2088834847	0.00261991846	0.00039791196	0.035983	0.28469

A_0 gen = [100], mem = [2, 2], pcs = 0.2,
A_1 gen = [100], mem = [2, 2], pcs = 0.3,
A_2 gen = [100], mem = [2, 2], pcs = 0.4
A_3 gen = [100], mem = [2, 2], pcs = 0.5,
A_4 gen = [100], mem = [3, 3], pcs = 0.2,
A_5 gen = [100], mem = [3, 3], pcs = 0.4
A_6 gen = [100], mem = [3,3], pcs = 0.5

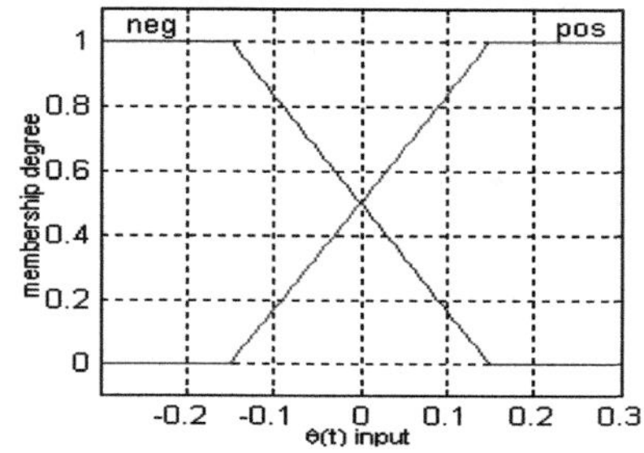

Figure 7.24 (a) Input membership functions for $\theta(t)$.

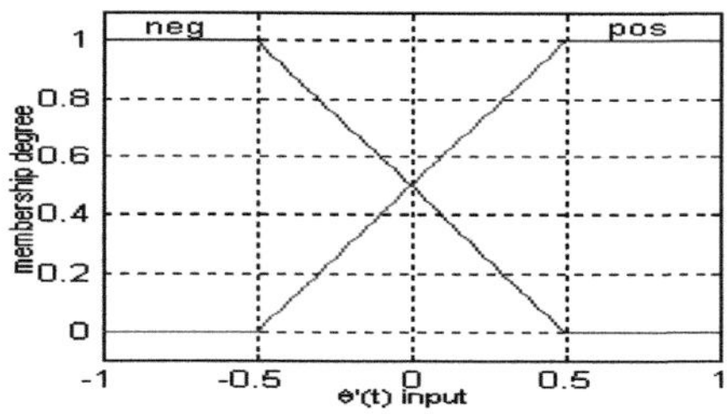

Figure 7.24 (b) Input membership functions for $\dot{\theta}(t)$.

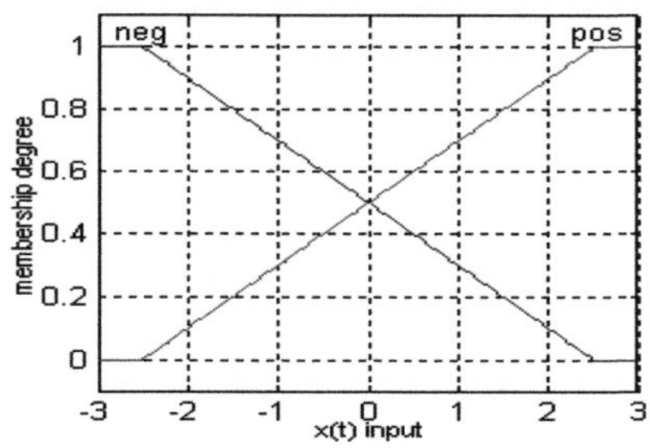

Figure 7.24 (c) Membership functions for $x(t)$.

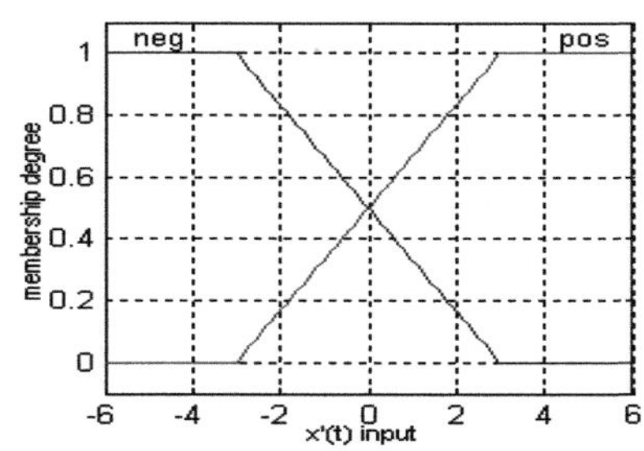

Figure 7.24(d) Input membership functions for $\dot{x}(t)$.

Table 7.4 Computed weight function by AIFNS

	w_1	w_2	w_3	w_4	w_5	w_6	w_7	w_8	e_i
1	1.7092	1.1366	1.8172	1.8651	0.9971	0.3958	0.4252	0.1375	2747.3
2	1.6256	1.3709	1.864	1.8614	1.1668	1.67	1.7432	0.2961	22.206
3	0.2030	0.6786	0.0061	0.7707	0.0365	0.3096	1.9885	0.4221	4.2395
4	0.9361	0.4963	1.2069	0.4873	0.7342	1.9955	1.7916	0.7562	0.6114
5	1.2857	1.2261	1.9534	1.9603	0.9868	1.7432	1.8806	0.5040	78.815
Φ: 5.0299									

Table 7.5 Fuzzy rules for position

	x	pos	pos
θ	$\dot{\theta}$	neg	pos
neg	neg	f_1	f_2
neg	neg	f_3	f_4
pos	pos	f_5	f_6
pos	pos	f_7	f_8

The AINFS-based generating membership shape of the fuzzy rules used in our Sugeno fuzzy controller are given in Table 7.5 and computed weight function in Table 7.6.

Table 7.6 Computed weight function by AINFS

	w_1	w_2	w_3	w_4	w_5	w_6	w_7	w_8	e_i
1	1.5573	1.3378	1.8236	1.9768	1.2524	0.4730	0.7890	0.5098	2565.1
2	1.4217	1.4463	1.9638	1.9708	1.0641	1.741	1.8501	0.3886	22.515
⋮	⋮	⋮	⋮	⋮	⋮	⋮	⋮	⋮	⋮
9	1.9562	1.6739	0.4960	0.6976	0.6808	1.652	0.7104	0.1879	2.8855
10	0.6437	1.2815	1.5168	1.5599	0.8642	1.6509	0.9751	0.5104	7.0122
Φ: 3.2245									

Figure 7.25 gives the response of angle control for θ and $\dot{\theta}$, while Fig. 7.26 shows the response of position control for x, $\dot{x}$.

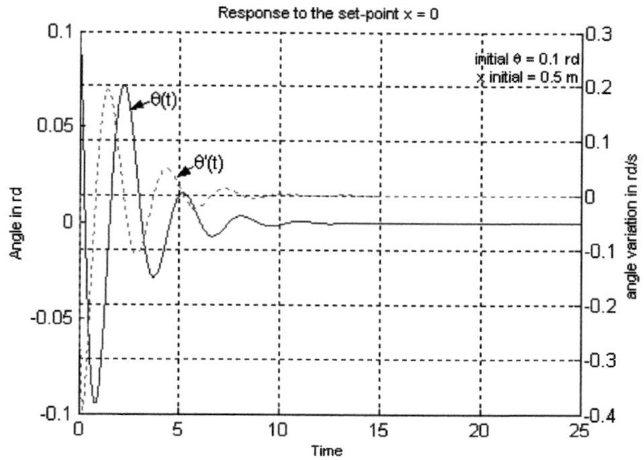

Figure 7.25 Response of angle.

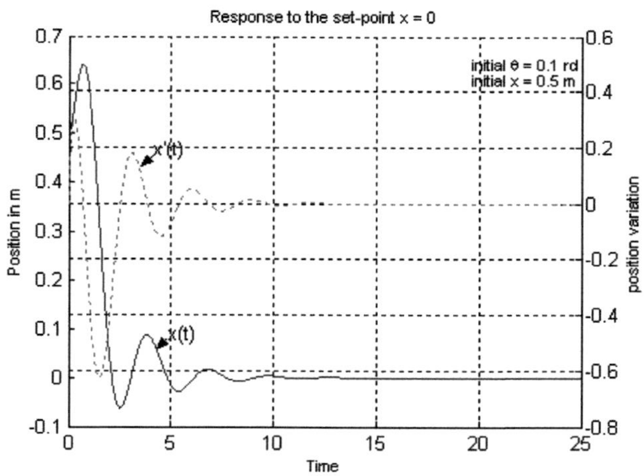

Figure 7.26 Response of position.

7.5.2.4 Response to variation *a*

Figures 7.27–7.29 show the responses obtained for various a. The values are also summarized in Table 7.7. Results obtained for various different values of a are obtained, to allow the study of the adaptation capability of the proposed AINFS controller. The corresponding adaptation characteristics are illustrated in Figs. 7.30–7.32 for θ and x as functions of α.

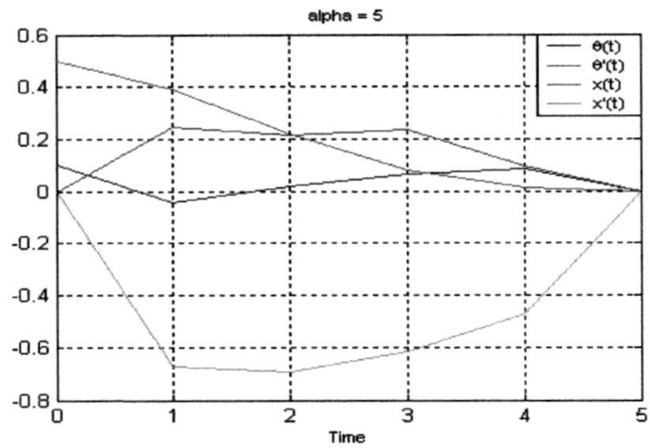

Figure 7.27 Response to variation $\alpha = 5$.

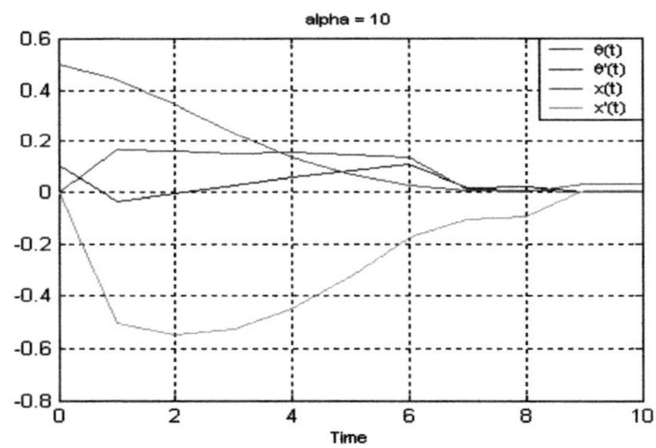

Figure 7.28 Response to variation $\alpha = 10$.

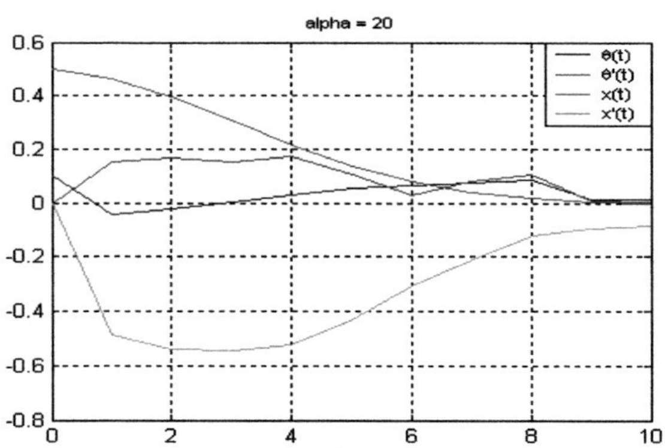

Figure 7.29 Response to antibody $\alpha = 20$.

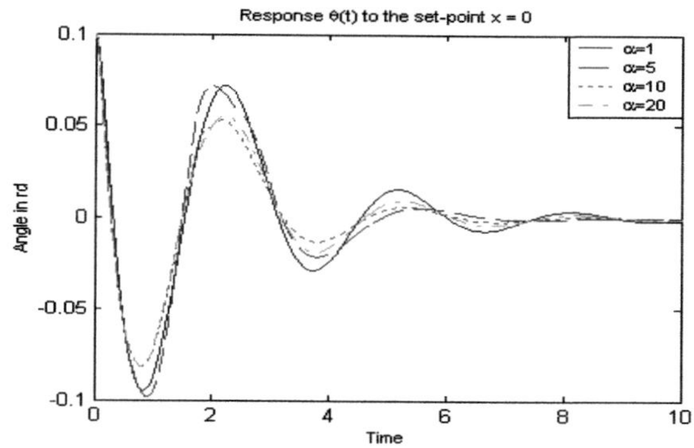

Figure 7.30 Variation of θ to α.

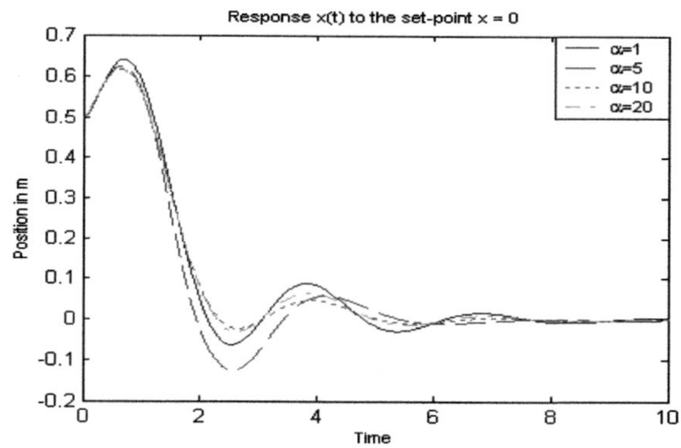

Figure 7.31 Variation of x to α.

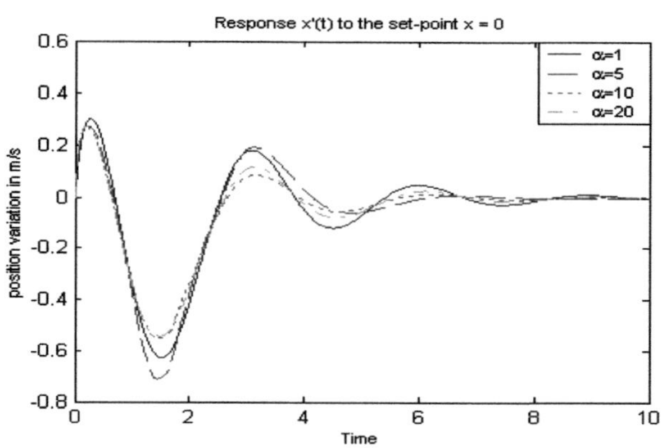

Figure 7.32 Variation of $\dot{x}$ to α.

Table 7.7 Computed weight function

	w_1	w_2	w_3	w_4	w_5	w_6	w_7	w_8	e_i
1	1.7092	1.1366	1.8172	1.8651	0.9971	0.3958	0.4252	0.1375	2747.3
2	1.6256	1.3709	1.864	1.8614	1.1668	1.67	1.7432	0.2961	22.206
3	0.2030	0.6786	0.0061	0.7707	0.0365	0.3096	1.9885	0.4221	4.2395
4	0.9361	0.4963	1.2069	0.4873	0.7342	1.9955	1.7916	0.7562	0.6114
5	1.2857	1.2261	1.9534	1.9603	0.9868	1.7432	1.8806	0.5040	78.815
Φ: 5.0299									

Obviously, the number of the antibodies is strongly related to the optimal gains acquired. Moreover, we apply the initial value 0.5 m and set point value 0 m to 1 m after 15 sec to obtain the optimal gains. When the set point changes between 0 and 1, the number of the antibodies depends on the characteristics of α while the number of the memory cells may significantly affect the control results. Tables 7.6 and 7.7 show relationship between the number of antibody and w of the Sugeno, in Figure 7.22. In summary, it is clearly visible that our adaptive fuzzy-immune control method can yield superior performance over the conventional controllers.

Figure 7.33 represents the fuzzy controller response at the initial point (position) x and $\dot{x}$, while Fig. 7.34 shows the response of fuzzy controller at initial point (angle) θ, $\dot{\theta}$. Figures 7.35 and 7.36 illustrate the variations of θ and $\dot{\theta}$ to α respectively. Figures 7.37 and 7.38 showcase the variations of x and $\dot{x}$ to α, respectively.

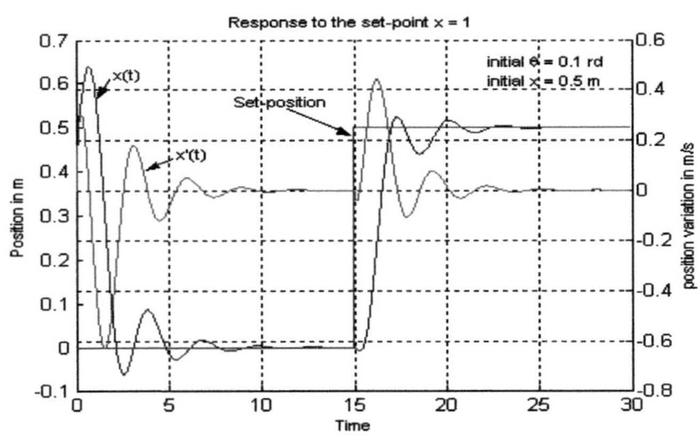

Figure 7.33 The response of AINFS controller at initial point (position).

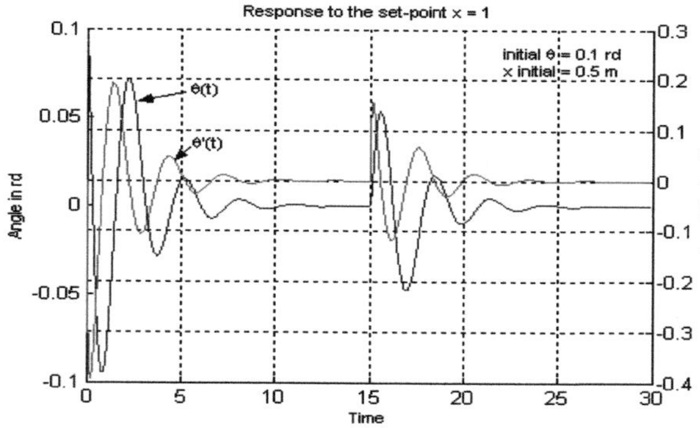

Figure 7.34 The response of AIFNS controller at initial point (angle).

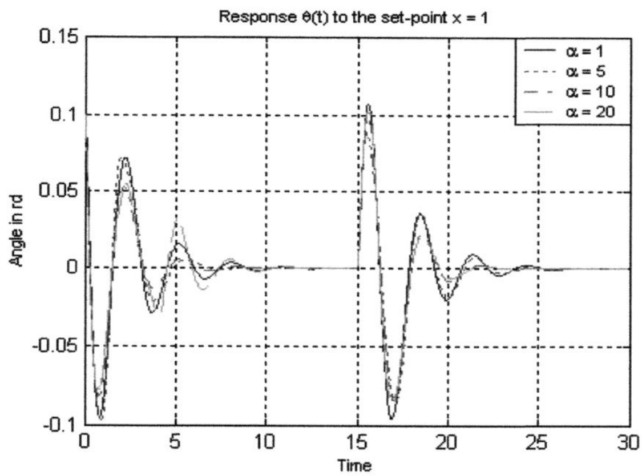

Figure 7.35 Variation of θ to α.

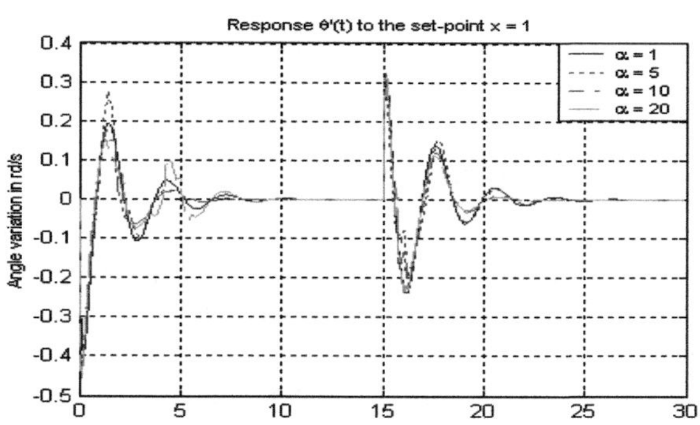

Figure 7.36 Variation of $\dot{\theta}$ to α.

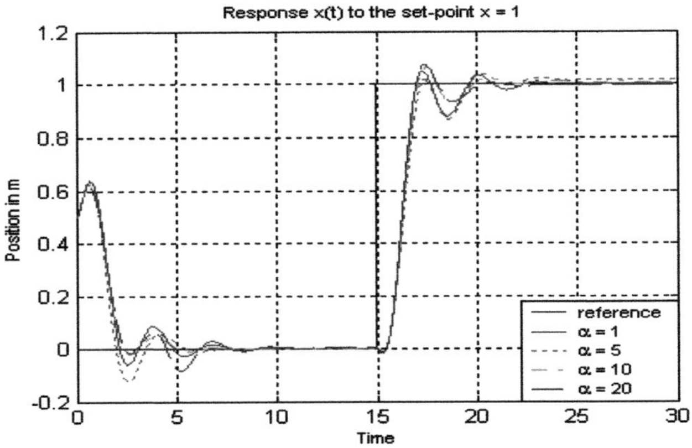

Figure 7.37 Variation of x to α.

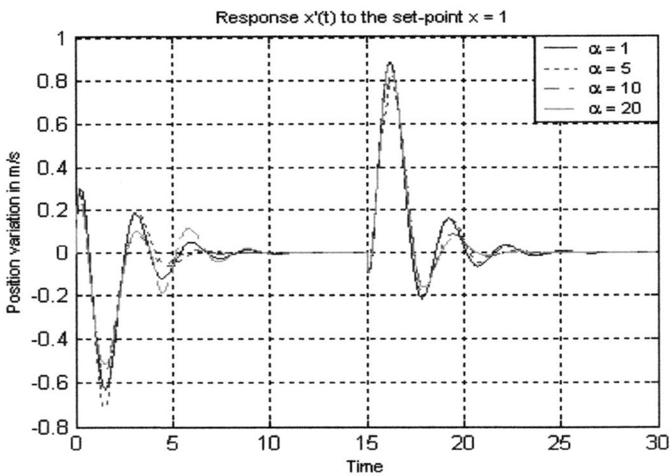

Figure 7.38 Variation of $\dot{x}$ with regard to α.

7.5.2.5 AINFS and discussions

It is well known that traditional controllers with fixed gains usually have difficulties in handling nonlinear plants with dynamically changing characteristics. We propose a novel nonlinear optimal control method called AIFNS. In our approach, the control characteristics are partitioned according to the nonlinearity defined by the fuzzy logic. Each partition is regulated by different optimal gains obtained using the immune algorithm whenever the system dynamics change. Optimal gains for the control sections are acquired based on the characteristics of the antibodies varying against their circumstances in the immune network. Simulations results obtained from applying the proposed control approach in the pendulum system have demonstrated its effectiveness in achieving enhanced performances.

7.6 Hybrid System (AINFNN) of ANIS and FNN for Nonlinear Gas Furnace System

7.6.1 The Structure of AINFNN Hybrid System

This chapter proposed an adjusting method of fuzzy membership function in FNNS (Fuzzy Neural Network System) by ANIS. The structure of the fuzzy-neural network system (FNNS) is shown in Fig. 7.39 [14] and the output of this FNNS can be represented by the following equation:

$$\begin{aligned}\bar{y} &= f_1(x_1) + f_2(x_2) + \dots + f_m(x_m) \\ &= \sum_{i=1}^{m} f_i(x_i)\end{aligned} \tag{7.16}$$

The output of the fuzzy neural network can be represented by the following simple equation:

$$\begin{aligned}f_i(x_i) &= \sum_{i=1}^{n} \mu_{xi} \cdot C_{yxi} \\ &= \frac{\sum_{j=1}^{n} \mu_{ij}(x_i) w_{ij}}{\sum_{j=1}^{n} \mu_{ij}(x_i)} = \frac{\mu_{ik}(x_i) w_{ik} + \mu_{ik+1}(x_i) w_{ik+1}}{\mu_{ik}(x_i) + \mu_{i,k+1}(x_i)} \\ &= \sum_{i=1}^{n} \mu_{xi} \cdot w_{xi}\end{aligned} \tag{7.17}$$

In this equation, the weights w_{ij} are assigned by learning the rule which is described by n “if-then” rules. That is, If input x_i lies in the fuzzy segment, μ_{ij}, then the corresponding weight w_{ij}, should be increased directly proportional to the output error $(y - \bar{y})$, because the error is caused by the weight. This gives

$$f_i(x_i) = \mu_{xi}(x_i)w_{xi} + \mu_{xi+1}(x_i)w_{xi+1} \tag{7.18}$$

In this chapter, the membership of this FNNS is tuned and applied for the nonlinear problem of a gas furnace.

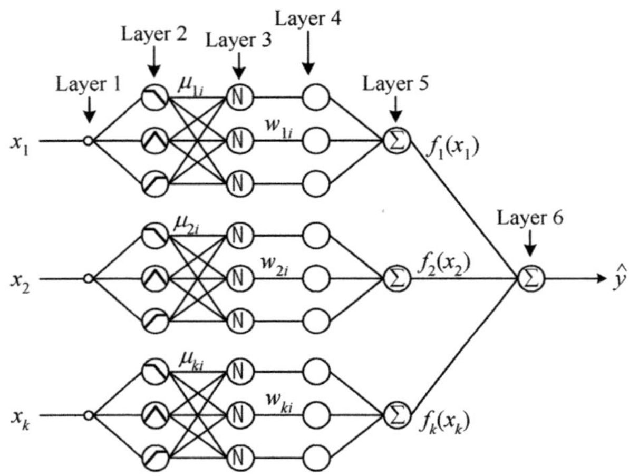

Figure 7.39 The structure of fuzzy-neural network for AINFNN.

7.6.2 Hybrid System of AINS and FNNS for Nonlinear Gas Furnace System

For proving the learning effect of the proposed AINS-based hybrid system, the second-order highly nonlinear difference equation (7.19) in [7.14] has been used:

$$y_k = \frac{y_{k-1}y_{k-2}(y_{k-1} + 2.5)}{1 + y_{k-1}^2 + y_{k-2}^2} + u_k \tag{7.19}$$

In the gas furnace equation (7.19), the values u $(t - 3)$ and y $(t - 1)$ are used as input, and $y(t)$ as output. When the initial value of the membership function is of triangular type (see Fig. 7.22), this is given by X_1_min = [0.46, 0.48], X_1_max = [0.77, 0.81], X_2_min = [45.0, 47.0], X_2_max = [61.0, 63.0]. One also has the learning rate boundary δ = [0.001, 0.01], momentum constant boundary

α = [0.00001, 0.0004], respectively. The final membership function obtained by immune algorithm is the dashed line shown in Figs. 7.40a,b.

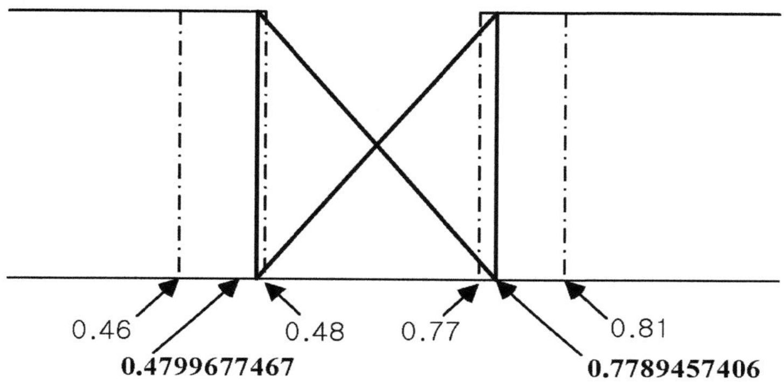

Figure 7.40 (a) Membership function shape of x1.

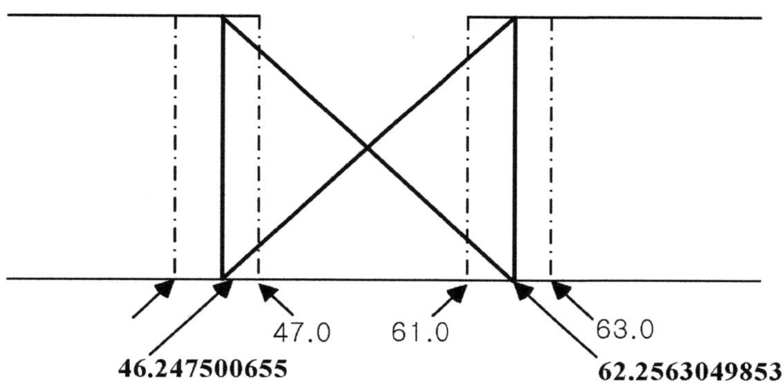

Figure 7.40 (b) Membership function shape of x1.

7.6.3 AINFNN-Based Computational Procedure

In this algorithm, we use the AINFNN-based calculation procedure from step 1 to 7 shown in Fig. 7.41 to optimize the learning rate, momentum term and fuzzy membership function of the above AINFS example. We use 10 and 100 generations, 60 populations, 10 bits per string, crossover rate equal to 0.6, and mutation probability equal to 0.1, respectively.

[Step 1] Initialization and recognition of antigen: The immune system recognizes the invasion of an antigen.

[Step 2] Product of antibody from memory cell: The immune system produces the antibodies that were effective to kill the antigen in the past. This is implemented by recalling a past successful solution from the memory cell. For each individual of the network population, calculate the fitness function using memory cell to membership function, learning rate and momentum constant.

[Step 3] The antibody with the best fitness value obtained by calculation for searching an optimal solution is stored in the memory cell.

[Step 4] Differentiation of lymphocyte: The B-lymphocyte cell, the antibody that matched the antigen, is dispersed to the memory cells in order to respond to the next invasion quickly. That is, select individuals using tournament selection and apply genetic operators (crossover and mutation) to the individuals of network.

[Step 5] Stimulation and suppression of antibody: The expected value ηk of the stimulation of the antibody is given by

$$\eta_k = \frac{m_{\varphi k}}{\sigma_k}, \tag{7.20}$$

where σ_k is the concentration of the antibodies.

An immune system can control the concentration and the variety of antibodies in the lymphocyte population. If antibody obtains a higher affinity against an antigen, the antibody stimulates. However, an excessively high concentration of an antibody is suppressed. Through this function, an immune system can maintain the diversity of searching directions and a local minimum.

[Step 6] Calculate fitness value between antibody and antigen. This procedure can generate a diversity of antibodies by a genetic reproduction operators such as mutation or crossover. These genetic operators are expected to be more efficient than the generation of antibodies.

[Step 7] If the maximum number of generations of memory cell is reached, stop and return the fitness of the best individual fitness value to network; otherwise, go to step 3.

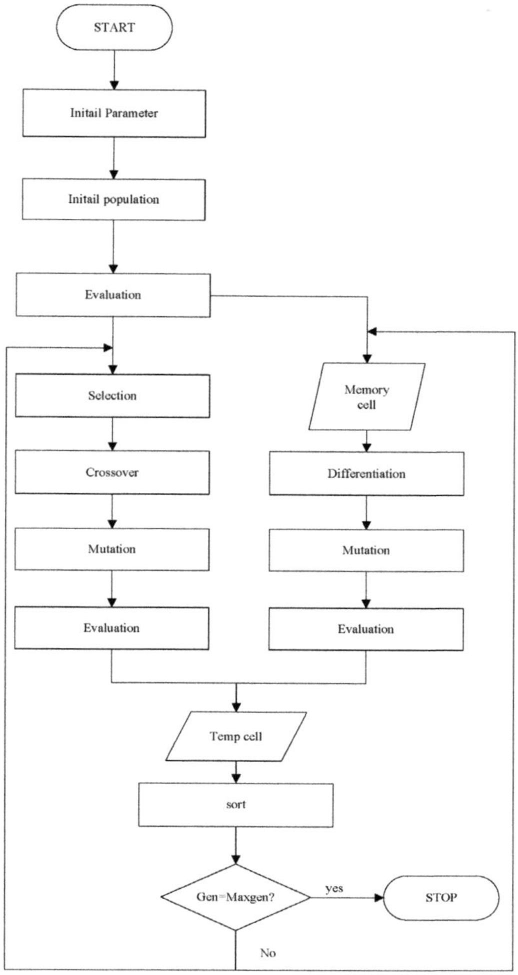

Figures 7.41 Computation procedure by clonal selection of the AINFNN.

7.6.4 Simulation and Discussions in Gas Furnace

Figure 7.42 shows the performance index by clonal differentiation rate of AINS for the given model (7.19) while Fig. 7.44 illustrates the best fitness and object functions depending on the differentiation rate of clonal selection pCS = 0.2. The number of membership function is 2.

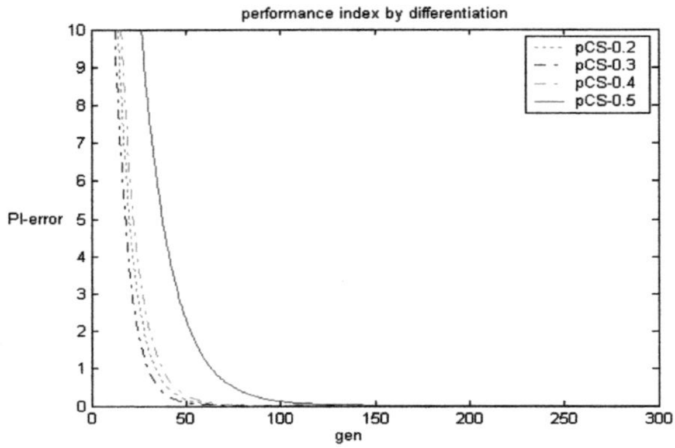

Figure 7.42 Performance index by differentiation rate of AINS.

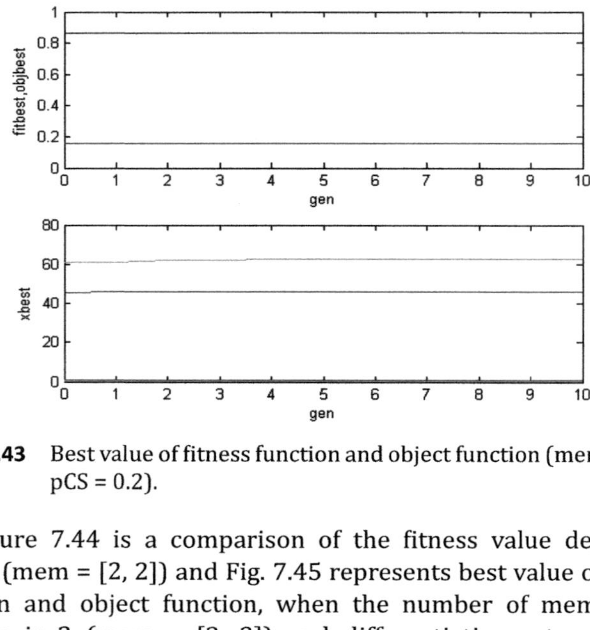

Figure 7.43 Best value of fitness function and object function (mem = [2, 2], pCS = 0.2).

Figure 7.44 is a comparison of the fitness value depending on pCS (mem = [2, 2]) and Fig. 7.45 represents best value of fitness function and object function, when the number of membership function is 3 (mem = [3, 3]) and differentiation rate of clonal pCS = 0.2, respectively. Figure 7.44 compares the fitness value by pCS to the number of membership function 3. Figures 7.45 and 7.46 show the best value of fitness when the learning parameter of the immune algorithm is 100 generations, 0.2 pCS (differentiation rate of clonal selection), and the number of membership is 2 and 3 respectively.

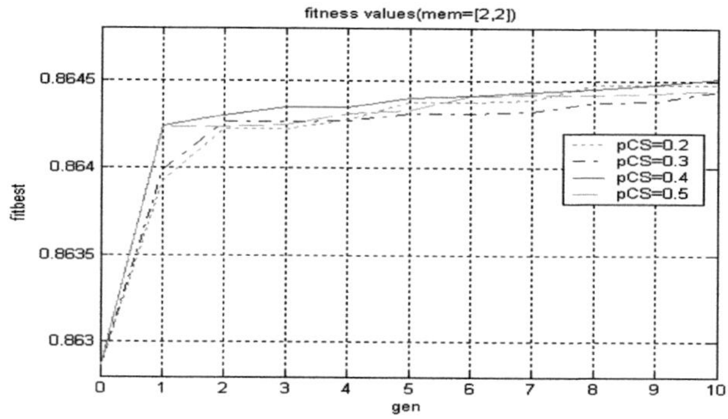

Figure 7.44 Comparison of fitness value depending pCS (mem = [2, 2]).

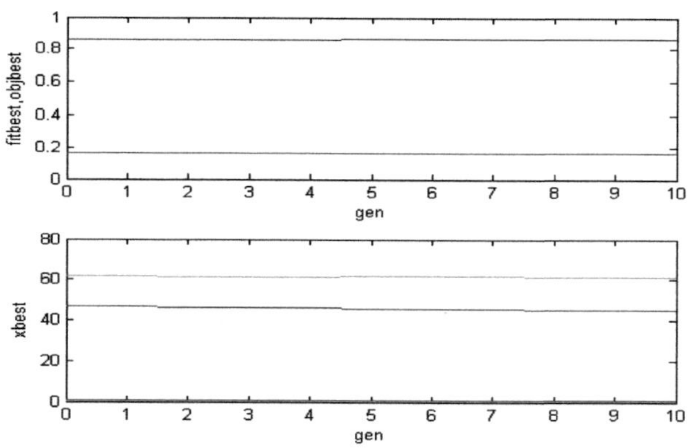

Figure 7.45 Best value of fitness function and object function (mem = [3, 3], pCS = 0.2).

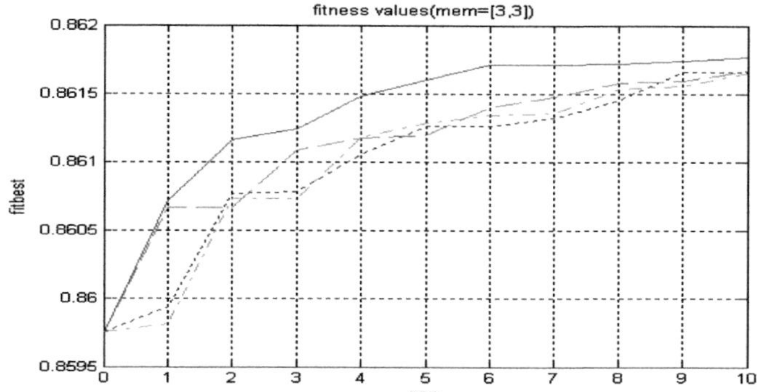

Figure 7.46 Comparison of fitness value by pCS (mem = [3, 3]).

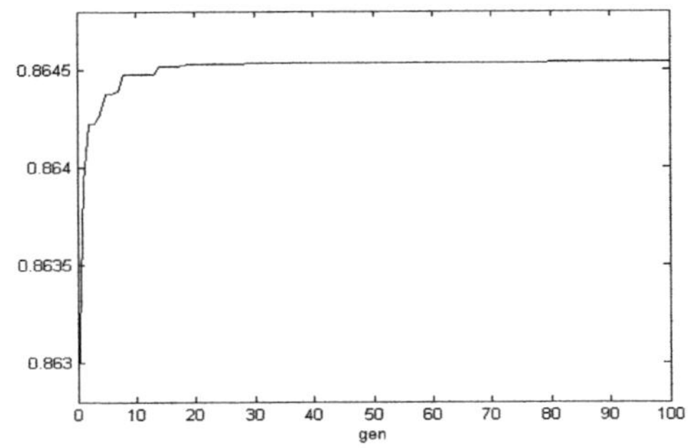

Figure 7.47 Best value of fitness (gen = 100, pCS = 0.2, mem = [2, 2]).

Figure 7.48 shows the best value of fitness by pCS (mem = [3, 3]), Figs. 7.49 and 7.50 are PI-error by pCS (mem = [2, 2]) and (mem = [3, 3]), and Figs. 7.51 and 7.52 show E_PI-error by pCS (mem = [2, 2]) and (mem = [3, 3]), respectively.

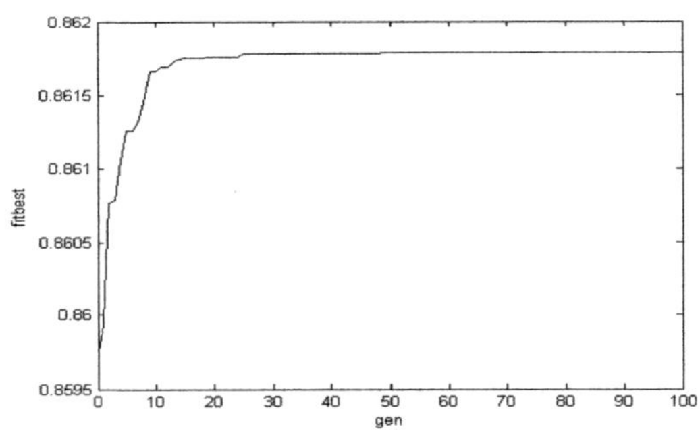

Figure 7.48 Best value of fitness (gen = 100, PCS = 0.2, mem = [3, 3]).

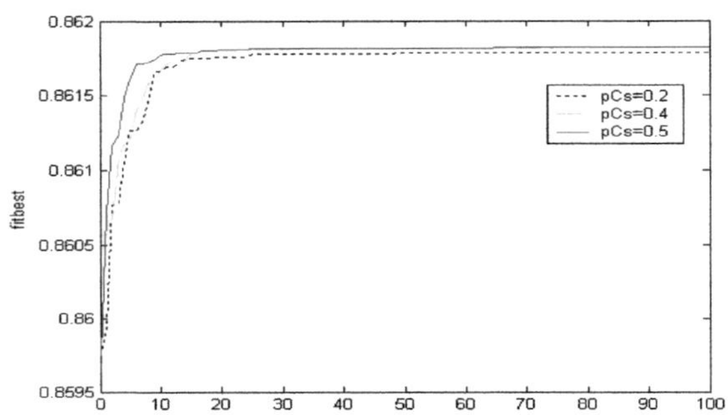

Figure 7.49 Best value of fitness by pCS (mem = [3, 3]).

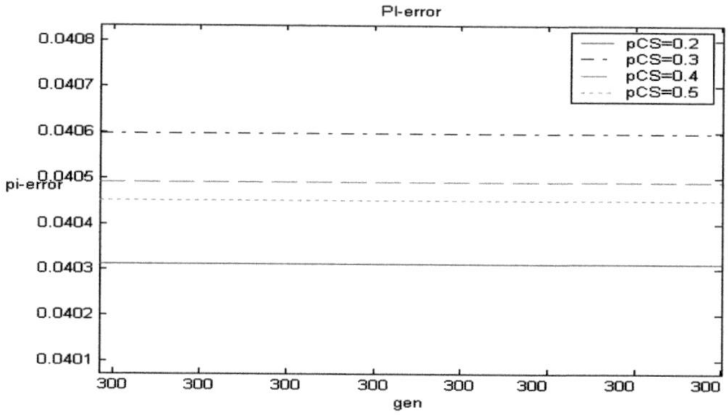

Figure 7.50 PI-error by pCS (mem = [2, 2]).

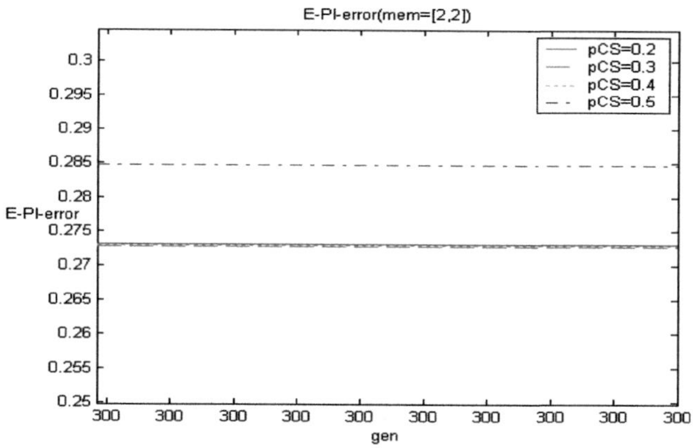

Figure 7.51 E_PI-error by pCS (mem = [2, 2]).

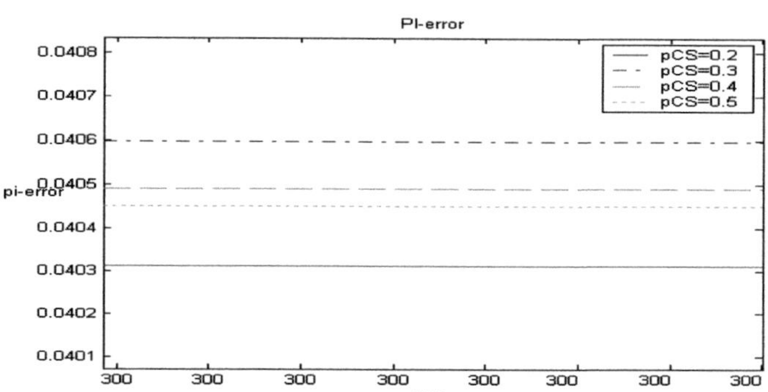

Figure 7.52 PI-error by pCS (mem = [3, 3]).

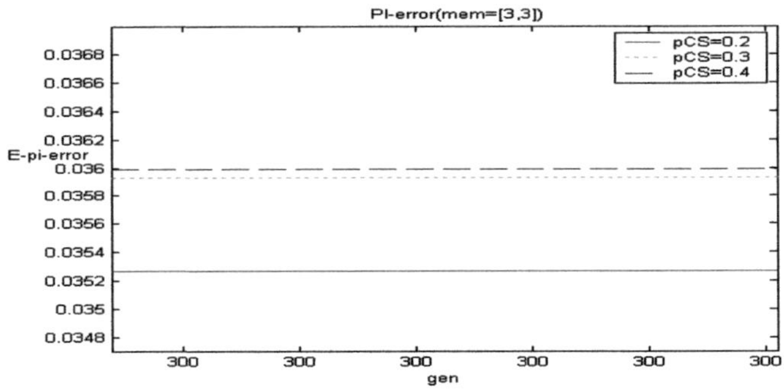

Figure 7.53 E_PI-error by pCS (mem = [3, 3]).

Table 7.8 shows the values of PI and E_PI by pCS for generations 10 and 100. That is, Table 7.8 is the shape of the generation dependent membership function of AINFNN. Table 7.9 depicts a comparison between the learning results obtained by a GA-based FNN model, HCM and GA-based FNN, and the AINS-based FNN model proposed in this chapter.

Table 7.8 Comparison of learning methods in 10 generation and 100 generation

	Gen = 10				Gen = 100			
pCS	2:2		3:3		2:2		3:3	
	PI	E_PI	PI	E_PI	PI	E_PI	PI	E_PI
0.2	0.0354	0.2857	0.0354	0.2857	0.040311	0.27306	0.035265	0.28551
0.3	0.0408	0.2729	0.0356	0.2855	0.040598	0.27277		
0.4	0.0409	0.2726	0.0359	0.2852	0.040491	0.27288	0.035923	0.28475
0.5	0.0394	0.2742	0.0361	0.2847	0.040452	0.27292	0.035983	0.28469

Table 7.9 also illustrates characteristics of various optimal learning methods. The proposed learning model described in this section is the FNN from which we can obtain optimized linguistic knowledge by AINS. The learning algorithm of an AINS-based FNNS (AINS-FNNS) is composed of finding the initialization of membership functions and then tuning of membership functions. The results are compared with the results by GA (genetic algorithm–based neural network), FNN, AINFNN respectively. The result for the learning method proposed in this chapter shows better values than the ones produced by other learning schemes.

Table 7.9 Comparison of the results by learning methods

Model	PI	E_PI	MF
FNN model (GA)	0.027	0.298	4
	0.026	0.304	6
FNN model (HCM+GA)	0.027	0.294	4
	0.032	0.276	6
FNN model (CS-GA)	0.0394	0.274	4
	0.0361	0.284	6

7.7 Conclusion

This session suggests a hybrid learning system by AINFS (Immune and Fuzzy) and AINFS-FNNS (Immune and Fuzzy neural network) for the optimal control of nonlinear systems and applies to level control systems such as car with pendulum and gas furnace.

The hybrid system tunes the shape of fuzzy membership function (by concentration of antibody in AINFS) to optimally control and reflect the characteristics of the nonlinear system. The AINFS hybrid system suggested in this session expressed good characteristics robustness when dealing with changing environment in nonlinear systems as shown in the previous session. Since fuzzy logic can capture the approximate, qualitative aspects of human reasoning and decision-making processes, they have been considered as effective tools to deal with uncertainties in terms of vagueness, ignorance, and imprecision.

On the other hand, neural networks (NN) appeared as designing high performance control systems, because they have the potential for dealing with favorable scenarios owing to nonlinear dynamics, drift in plant parameters, and shifts in operating points.

The fuzzy-neural network system (FNNS) learning represents one of the most effective existing algorithms used to build such linguistic models for control systems or decision making.

The AINFNN is a convenient tool for artificial intelligence because of advantages such as: no leader processing, distribution function, or networking function between antibodies in information processing. Generally, in many cases, tuning of membership and weighting function remains a difficult task. Some papers suggested that genetic algorithms (GAs) may be the only efficient and robust tools for generating fuzzy rules and weighting functions. However, the hybrid intelligent system suggested in this chapter provides a viable alternative, because optimal behavior is achieved by selecting the advantages of each algorithm. It can also be used for nonlinear system or robust control.

Acknowledgments

This research work was supported by the Basic Science Research Program through the National Research Foundation of Korea (NRF) funded by the Ministry of Education, Science and Technology (D00046: Bio based Intelligent Computing Technology) and (2012042098: 2012R1A1A2042098). This research was also supported by the research fund of Hanbat National University for supporting of research professor in 2012 (Research in Hungary), combining department (Bio based intelligent emotion pattern recognition) in 2013, and the research fund of Hanbat National University in 2013.

References

1. Y.-Y. Cao, Stability analysis and synthesis of nonlinear time-delay systems via linear Takagi-Sugeno fuzzy models, *Fuzzy Sets Syst.*, vol. 124, pp. 213–229, 2001.
2. C.-S. Ting, T.-H. S. Li, F.-C. Kung, An approach to systematic of the fuzzy control system, *Fuzzy Sets Syst.*, vol. 77, pp. 151–166, 1996.
3. M. Margaliot, G. Langholz, Fuzzy lynapunov-based approach to the design of fuzzy controllers, *Fuzzy Sets Syst.*, vol. 106, pp. 49–59, 1999.
4. Homaifar, E. Mccormick, Simultaneous design of membership functions and rule sets for fuzzy controllers using genetic algorithms, *IEEE Trans. Fuzzy Syst.*, vol. 3, pp. 129–139, 1995.
5. C. Lee, Fuzzy logic in control system: Fuzzy logic controller, part I and II, *IEEE Trans. Syst. Man Cybern*, vol. 20, no. 2, pp. 404–435, 1990.
6. D. H. Kim, J. H. Cho, Robust tuning of PID controller using bacterial-foraging-based optimization, *J. Adv. Comput. Intell. Intell. Inform.*, vol. 9(6), pp. 669–676, 2005.
7. R. Brooks, Intelligence without reason, *Proceedings of the IJCAI-91*, pp. 569–595, 1991.
8. Ishiguro, T. Kondo, Y. Watanabe, Y. Uchikawa, Dynamic behavior arbitration of autonomous mobile robots using immune networks, *Proc. ICEC' 95*, vo1. 2, pp. 722–727, 1995.
9. J. Park, I. W. Sandberg, Approximation and radial-basis-function networks, *Neural Comput.*, pp. 105–110.
10. H. Wang, M. Brown, and C. J. Harris, Neural network modeling of unknown nonlinear systems subject to immeasurable disturbances, *IEE Proc. Control Theory Appl.*, vol. 141, no. 4, pp. 216–222, 1994.
11. S. Horikawa, T. Furuhashl, and Y. Uchikawa, On fuzzy modeling using fuzzy neural networks with back propagation algorithm, *IEEE Trans. Neural Netw.*, vol. 3, no. 7, pp. 801–806, 1992.
12. Wael, V. H. Farag, Quintana, and Germano Lambert-Torred, A genetic-based neuro-fuzzy approach for modeling and control of dynamical systems, *IEEE Trans. Neural Netw.*, vol. 9, no. 5, pp. 607–704, Sept. 1998.

13. Y. C. Chen and C. C. Teng, Fuzzy neural network systems in model reference control systems, in *Neural Network Systems, Technique and Applications* (C. T. Leondes, ed.), vol. 6, Academic Press, Inc. pp. 285–313, 1998.
14. S. Matsummura, Adaptive control for the steam temperature of thermal power plants, *Proceedings the 1993 IEEE on Control Applications*, pp. 1105–1109, Sept. 1998.
15. A. Blazina, Neural network-based feedforward control of two-stage heat exchange process, *IEEE conference*, pp. 25–29, 1997.
16. H. Kim, Application of a multivariable PID controller with a neural network tuning method to the heat exchange, *FUZZ-IEEE* (Seoul), pp. 23–25, Aug. 1998.
17. Y. Z. Lu, *Industrial Intelligent Control*, John Wiley & Sons, 1996.
18. H. Kim, Aapplication of intelligent control algorithms, *Conference of ICASE* (Seoul), pp. 15–17, 1997.
19. D. Lindsley, *Boiler Control Systems*, McGrawill, 1991.
20. F.-C. Teng, Self-tuning PID controllers for dead time process, *IEEE Trans.*, vol. 35, no. 1, pp. 119–125, 1988.
21. Y.-G. Wang, PI tuning for processes with dead time, *AACC2000*, Chicago, Illinois, June, 2000.
22. H.-J. Cho, K.-B. Cho, B.-H. Wang, Automatic rule generation using genetic algorithms, *Fuzzy Sets Syst.*, vol. 92, pp. 305–316, 1997.
23. C.-H. Lee, C.-C. Ten, Calculation of PID controller parameters by using a fuzzy neural network, *ISA Trans.*, vol. 42, no. 3, pp. 391–400, 2003.
24. G. K. I. Mann, B.-G. Hu, R. G. Gosine, Analysis of direction fuzzy PID controller structures, *IEEE Trans. Syst. Man Cybern. Part B*, vol. 29, no. 3, pp. 371–388, 1999.
25. B. Stuart, Development of PID controller, *IEEE Control Syst.*, vol. 13, no. 6, pp. 58–62, 1993.
26. Y. Stephen, A laboratory course on fuzzy control, *IEEE Trans. Educ.*, vol. 42, no. 1, pp. 15–21, 1999.
27. W. K. Ho, PID tuning for unstable process based on gain and phase-margin specifications, *IEEE Proc. Control. Theory Appl.*, vol. 45, no. 5, pp. 392–396, 1998.
28. R. Ketata, D. De Geest, and A. Titli, Fuzzy controller: Design, evaluation, parallel and hierarchical combination with a PID controller, *Fuzzy Sets Syst.*, vol. 71, pp. 113–129, 1995.

29. I.-L. Chien and P. S. Fruehauf, Consider IMC tuning to improve controller performance, *Chem. Eng. Prog.*, vol. 86, no. 10, pp. 33–41, 1990.

30. C.-L. Lin, H.-W. Su, Intelligent control theory in guidance and control system design: An Overview, *Proc. Natl. Sci. Counc. ROC (A)*, vol. 24, no. 1, pp. 15–30, 2000.

31. E. Figueres, J.-M. Benavent, G. Garcera, Robust control of power-factor-correction rectifiers with fast dynamic response, *IEEE Trans. Ind. Electron.*, vol. 52, no. 1, pp. 66–76, 2005.

32. J. Azcondo, C. Rranas, R. Casanueva, and S. Bracho, Power-mode-controlled power-factor corrector for electronic ballast, *IEEE Trans. Ind. Electron.*, vol. 52, no. 1, 2005.

33. K. J. Astrom and T. Hagglund, Automatic tuning of simple regulators with specifications on phase and amplitude margins, *Automatica*, vol. 20, no. 5, pp. 645–651, 1984.

34. C. H. Lee and C. C. Teng, A novel robust PID controllers design by fuzzy neural network, *Asian J. Control.*, vol. 4, no. 4, pp. 433–438, 2002.

35. J. X. Xu, Y. M. Pok, C. Liu, and C. C. Hang, Tuning and analysis of a fuzzy PI controller based on gain and phase margins, *IEEE Trans. Syst. Man Cybern. Part A Syst. Hum.*, vol. 28, no. 5, pp. 685–691, 1998.

36. T. W. Kraus and T. J. Mayron, Self-tuning PID controllers based on a pattern recognition approach, *Control. Eng. Pract.*, 106–111, 1984.

37. C. C. Hang, W. K. Ho, and L. S. Cao, A comparison of two design methods for PID controllers, *ISA Trans.*, 33, 147–151, 1994.

38. M. Zhuang and D. P. Atherton, Automatic tuning of optimum PID controllers, *IEEE Proc. Part D*, 14, pp. 216–224, 1993.

39. J. G. Ziegler and N. B. Nichols, Optimum settings for automatic controllers, *Trans. ASME*, 64, pp. 759–768, 1942.

40. C. H. Lee and C. C. Teng, Tuning PID controller of unstable processes: A fuzzy neural network approach, *Fuzzy Sets Syst.*, vol. 128, no.1, pp. 95–106, 2002.

41. K. J. Astrom, B. Witternmark, *Adaptive Control*, Addison-Wesley Publishing Com., 1995.

42. J. D. Farmer, N. H. Packard, and A. S. Perelson, The immune system, adaptation, and machine learning, *Phys. D*, no. 22, pp. 187–204, 1986.

43. K. Mori and M. Tsukiyama, Immune algorithm with searching diversity and its application to resource allocation problem, *Trans. JIEE*, vol. 113 C, no. 10, '93.

44. C. V. Rao, *An Introduction to Immunology*, Alpha Science International Ltd. 2002.

45. R. Brooks, A robust layered control system for a mobile Robot, *IEEE J. RA*, vol. 2, no. 3, pp. 14–23, 1986.

46. N. K. Jerne, The immune system, *Sci. Am.*, vo1. 229, no. 1, pp. 52–60, 1973.

47. N. K. Jerne, Idiotypic networks and other preconceived ideas, *Immunol. Rev.*, vol. 79, pp. 5–24, 1984.

48. J. Valera, A. Coutinho, B. Dupire, and N. N. Vaz., Cognitive networks: Immune, neural, and Otherwise, *Theor. Immunol.*, vo1. 2, pp. 359–375, 1988.

49. J. Stewart, The immune system: Emergent self-assertion in an autonomous network, *Proceedings of ECAL-93*, pp. 1012–1018, 1993.

50. J. D. Farmer, S. A. Kauffman, N. H. Packard, and A. S. Perelson, Adaptive Dynamic Networks as Models for the Immune System and Autocatalytic Sets, Technical Report LA-UR-86-3287, Los Alamos National Laboratory, Los Alamos, NM, 1986.

51. K. Nakano, H. Hiraki, and S. Ikeda, A learning machine that evolves, *Proc. ICEC-95*, pp. 808–813, 1995.

52. G. J. Nossal, Various Authors, Life, death and the immune system, *Sci. Am.*, vol. 269, no.3, pp. 20–102, 1993.

53. S. Forrest, S. A. Hofmeyr, and A. Somayaji, Computer immunology, *Commun. ACM*, vol. 40, no. 10, pp. 88–96, 1997.

54. D. Gray, The dynamics of immunological memory, *Semin. Immunol.*, vol. 4, pp. 29–34, 1992.

55. W. D. Hamilton, R. Axelrod, and R. Tanese, Sexual reproduction as an adaptation to resist parasites, *Proc. Natl. Acad. Sci. U. S. A.*, 87, pp. 3566–3573, 1990.

56. J. K. Inman, The antibody combining region: Speculations on the hypothesis of general multispecificity, *Theor. Immunol.*, 1978.

57. C. A. Janeway and P. Travers, *The Immune System in Health and Disease, Immunobiology*, 2nd ed. Current Biology Ltd., London, 1996.

58. C. R. MacKay, Immunological memory, *Adv. Immunol.*, vol. 53, pp. 217–265, 1993.

59. S. Forrest, B. Javornik, R. E. Smith, and A. S. Perelson, Using genetic algorithms to explore pattern recognition in the immune system, *Evol. Comput.*, vol. 1, pp. 191–211, 1993.

60. F. D'Alche-Buc, V. Andres, and J-P. Nadal, Rule extraction with fuzzy neural network, *Int. J. Neural Syst.*, vol. 5, pp. 1–11, 1994.

61. D. H. Kim, Intelligent control of nonlinear dynamic systems using immune fuzzy fusion, *JACI*, vol. 7, no. 3, 2003.

62. D. H. Kim, Tuning of new 2-DOF PID controller for Gun-san gas turbine by adaptive neural fuzzy, *Inference Syst.*, *JACI*, vol. 5, no. 5, 2001.

63. D. H. Kim, Intelligent control of inverted pendulum system using immune fuzzy fusion, *WSEAS Trans. Comput.*, vol. 3, no. 3, pp. 552–557, July, 2004.

64. D. H. Kim, J. H. Cho, Robust PID using Gain/Phase margin and advanced immune algorithm, *WSEAS Trans. Syst.*, vol. 3, no. 9, pp. 2841–2851, 2004.

65. D. H. Kim, Tuning of a PID controller using a artificial immune network model and fuzzy set, *IFSA 2001* (Vancouver), July 28, 2001.

66. D. H. Kim, Neural networks control by immune algorithm based auto-weight function tuning, *2002 IEEE International Conference on Neural Networks* (Honolulu, Hawaii), May 12–16, 2002.

67. D. H. Kim, Auto-tuning of reference model based PID controller using immune algorithm, *2002 IEEE International Conference on Evolutionary Computation* (Honolulu, Hawaii), May 12–16, 2002.

68. D. H. Kim, J. H. Cho, Intelligent tuning of PID controller with Robust disturbance rejection function using immune Algorithm, *Lecture Notes in Computer Science Proceeding of Springer* (SCI) Sept. 22–24, 2004.

69. D. H. Kim, Neural network tuning of 2-DOF PID Controller with a combined 2-DOF parameter for nuclear steam generator level control, *Lecture Notes in Computer Science Proceeding of Springer* (SCI) Sept. 22–24, 2004.

70. D. H. Kim, J. H. Cho, Disturbance rejection control of thermal power plant using immune algorithm, *Lecture Notes in Computer Science Proceeding of Springer* (SCI) June 7–9, 2004.

71. D. H. Kim, PID controller tuning of a boiler control system using immune algorithm typed neural network, *Lecture Notes in Computer Science Proceeding of Springer* (SCI) June 7–9, 2004.

72. D. H. Kim, etc., The Genetic design of hybrid fuzzy controllers, *Cybern. Syst. An Int. J.* (SCI), vol. 35, no. 4, pp. 1–29, 2003.

73. D. H. Kim, Intelligent PID control by immune algorithm based fuzzy rule auto-tuning, *IFSA Proceeding* (SCI), June 29-July 2, 2003.

74. D. H. Kim, Intelligent control of nonlinear dynamic systems using immune fuzzy fusion, *JACI*, vol. 7, no. 3, pp. 330–338, December 2003.
75. D. H. Kim, Tuning of new 2-DOF PID controller for Gun-san gas turbine by adaptive neural fuzzy inference system, *JACI*, vol. 5, no. 5, April 2001.
76. D. H. Kim, Hwan Lee, Intelligent control of nonlinear power plant using immune algorithm based multiobjective optimization, *IEEE ICNSC2004* (Taipei), March 21–23, 2004.
77. D. H. Kim, Intelligent tuning of the 2-DOF PID controller with a separated 2-DOF parameter using immune network for Boiler control of thermal power plant, *IFAC* (Barocelona, Spain), July 21–26, 2002.
78. D. H. Kim, Adaptive controller design for multivariable system using immune algorithms, *IEEE CIRAS2003* (Singapore), Dec. 15–18 2003.
79. D. H. Kim, Intelligent tuning of dead time process using immune based multiobjective, *IASTED* (Canada), July 14–16, 2003.
80. D. H. Kim, Comparison of PID controller tuning of power plant using immune and genetic algorithms, *IEEE-CIMSA 2003* (Ligano, Switzerland), pp. 29–31, July 2003.
81. D. H. Kim, Temperature control of power plant using immune algorithm based decoupling computation, *SCIS&ISIS 2002* (Japan), Oct. 21–25, 2002.
82. D. H. Kim, An Artificial immune network algorithm based 2-DOF controller system design, *SCIS&ISIS 2002* (Japan), Oct. 21–25, 2002.
83. D. H. Kim, Intelligent tuning of the 2-DOF PID controller on the DCS for steam temperature control of thermal power plant, *IEEEI&CPS* (Savannah, USA), May 7, 2002.
84. B. J. Park, Fuzzy polynomial neural networks: Hybrid architectures of fuzzy modeling, *IEEE Trans. Fuzzy Syst.*, vol. 10, no. 5, pp. 607–621, Oct. 2002.
85. D. H. Kim, Robust PID controller tuning using multiobjective optimization based on clonal selection of immune algorithm, *Proceedings of the International Conference on Knowledge-Based Intelligent Information and Engineering Systems*. Springer-Verlag. pp. 50–56, 2004.
86. D. H. Kim, Intelligent control by immune algorithm based weight parameter tuning of fuzzy neural networks, *IEEE CIRAS2003* (Singapore), Dec. 15–18, 2003.
87. D. H. Kim, J. H. Cho, Disturbance rejection control of thermal power plant using immune algorithm, *Lecture Notes in Computer Science Proceeding of Springer* (SCI) June 7–9, 2004.

88. D. H. Kim, Robust intelligent tuning of PID controller for multivariable system using clonal selection and fuzzy logic, *Lecture Notes in Computer Science Proceeding of Springer* (SCI), June 12–15, KES 2005, Melbourne, Australia, 2005.

89. D. H. Kim, etc., The Genetic design of hybrid fuzzy controllers, *Cybern. Syst. Int. J.* (SCI), vol. 35, no. 4, pp. 1–29, 2003.

90. S.-K. Oh, D.-K. Lee, W. Pedrycz, D.-H. Kim, The genetic design of hybrid fuzzy controllers, *Cybern. Syst. Int. J.* (SCI), vol. 35, pp. 333–361, 2004.

91. D. H. Kim, Optimal leaning of fuzzy neural network using particle swarm optimization, *Lecture Notes in Computer Science Proceeding of Springer* (SCI), May 12–15, Atlanta, 2005.

92. D. H. Kim, J. Ill Park, Intelligent tuning of PID controller for AVR system using a hybrid GA-PSO approach, *Lecture Notes in Computer Science Proceeding of Springer* (SCI), May 12–15, Atlanta, 2005.

93. D. H. Kim, J. Ill Park, Loss minimization control of induction motor using GA-PSO, *Lecture Notes in Computer Science Proceeding of Springer* (SCI), Sept. 12–15, KES 2005, Melbourne, Australia, 2005.

94. D. H. Kim, J. Ill Park, Improvement of genetic algorithm using PSO and euclidean data distance, *IEEE 2005 Intelligent Computing, Lecture Notes in Computer Science Proceeding of Springer* (SCI), Aug. 23–26, China, 2005.

95. D. H. Kim, J. Ill Park, Intelligent PID controller tuning of AVR system using GA and PSO, *IEEE 2005 Intelligent computing, Lecture Notes in Computer Science Proceeding of Springer* (SCI), Aug. 23–26, China, 2005.

96. D. H. Kim, J. H. Cho, *Particle Swarm Optimization Algorithm Based Fuzzy Neural Network*, International Fuzzy System Association, Tsinghua University, Sichuan University, Aug. 28–31, 2005.

97. D. H. Kim, Robust tuning of PID controllers with disturbance rejection using bacterial foraging based optimization, *WSEAS Trans. Syst.*, vol. 3, no. 9, 2004, pp. 2834–2840.

98. D. H. Kim, Robust tuning of embedded intelligent PID controller for induction motor using bacterial foraging based optimization (Best paper candidate, *Lecture Notes in Computer Science Proceeding of Springer* (SCI)–LNCS, ICESS2004, Dec. 9–10, Zhejiang University, Hanzhou, China, 2004.

99. D. H. Kim, J. H. Cho, Adaptive tuning of PID controller for multivariable system using bacterial foraging based optimization, *Lecture Notes*

in Computer Science Proceeding of Springer (SCI), June 12–15, Poland, 2005.

100. D. H. Kim, J. H. Cho, Intelligent control of AVR system using GA-BF, *Lecture Notes in Computer Science Proceeding of Springer* (SCI), Sept. 12–15, KES 2005, Melbourne, Australia, 2005.
101. D. H. Kim, Decentralised PID controller tuning for multivarlable process using multiobjective optimization based on bacterial foraging IEEE 2005 intelligent computing, *Lecture Notes in Computer Science Proceeding of Springer* (SCI), Aug. 23–26, China, 2005.
102. Z. Michalewicz, Genetic Algorithms + Data Structures = Evolution Programs, New York: Springer-Verlag, 1999.
103. J. Arabas, Z. Michalewicz, and J. Mulawka, GAVaPS—A genetic algorithm with varying population size, *Proceedings of the IEEE Internationa Conference on Evolutionary Computation* (Orlando), pp. 73–78, 1994.
104. R. Tanese, Distributed genetic algorithm, *Proceedings of the International Conference on Genetic Algorithms*, pp. 434–439, 1989.
105. Y. Shi and R. Eberhart, A modified particle swarm optimizer, *Proceedings of the IEEE World Congress on Computational Intelligence*, pp. 69–73, May 1998.
106. Y. K. Kawata, and Y. Fukuyama, A particle swarm optimization for reactive power and voltage control considering voltage security assessment, *IEEE Trans. Power Syst.*, vol. 15, pp. 1232–1239, Nov. 2000.
107. C.-F. Juang, A hybrid of genetic algorithm and particle swarm optimization for recurrent network design, *IEEE Trans. Syst. Man Cybern. Part B*, vol. 34, pp. 997–1006, 2, April, 2004.
108. R. A. Krohling and J. P. Rey, Design of optimal disturbance rejection PID controllers using genetic algorithms, *IEEE Trans. Evol. Comput.*, vol. 5, pp. 78–82, Feb. 2001.
109. K. J. Astr¨om, T. Hagglund, C. C. Hang, and W. K. Ho, Automatic tuning and adaptation for PID controllers—A survey, *IFAC J. Contr. Eng. Pract.*, vol. 1, no. 4, pp. 699–714, 1993.
110. D. W. Stephens and J. R. Krebs, *Foraging Theory*, Princeton University Press, Princeton, New Jersey, 1986.
111. J. Alcock, *Animal Behavior, An Evolutionary Approach*, Sinauer Associates, Sunderland, Massachusetts, 1998.
112. W. J. Bell, *Searching Behavior, The Behavioral Ecology of Finding Resources*, Chapman and Hall, London, England, 1991.

113. D. Grunbaum, Schooling as a strategy for taxis in a noisy environment, *Evol. Ecol.*, vol. 12, pp. 503–522, 1998.

114. E. Bonabeau, M. Dorigo, and G. Theraulaz, *Swarm Intelligence: From Natural to Artificial Systems*, Oxford University Press, New York, NY, 1999.

115. K. M. Passino, *Biomimicry of Bacterial foraging for Distributed Optimization*, University Press, Princeton, New Jersey, 2001.

116. K. M. Passino, Biomimicry of bacterial foraging for distributed optimization and control, *IEEE Control Syst. Mag.*, vol. 22, pp. 52–67, 2002.

117. D. H. Kim, et al., Intelligent PID controller tuning using GA-BF, SOFT, Submitted material, Jan. 2005.

118. P. Marrack and J. W. Kappler, How the immune system recognizes the body, *Sci. Am.*, vol. 269, no. 3, pp. 102–108, 1993.

119. I. Y. Watanabe and Y. Uchikawa, An immunological approach to dynamic behavior control for autonomous mobile robots, *Proc. IROS'95*, vol. 1, pp. 495–500, 1995.

120. F. T. Vertosick and R. H. Kelly, Immune network theory: A role for parallel distributed processing, *Immunology*, vol. 66, pp. 1–7, 1989.

121. F. J. Varela and J. Stewart, Dynamics of a class of immune networks I. Global stability of idiotype interactions, *J. Theor. Biol.*, vol. 144, no. 1, pp. 93–101, 1990.

122. M. N. Lima, C. B. Jacobina, E. B. Souza Filho, Nonlinear parameter estimation of steady-state induction machine models, *IEEE Trans. Ind. Electron.*, vol. 44, no. 3, 1997.

123. J. Hickiewicz, K. Macek-Kaminska, and P. Wach, Algorithmic methods of induction machines parameters estimation from measured slip curves, *Arch. Elektrotech.*, vol. 72, Issue 4, pp. 239–249, 1989.

124. D. Lindsley, *Boiler Control Systems*, McGraw-Hill International Editions, 1991.

125. B.-Adbennpur, A decentralized controller design for a power plant using robust local controllers and functional mapping, *IEEE Trans. Energy Conversion*, vol. 11, no. 2, pp. 394–400, June 1996.

126. E. G. Kim, Commissioning results of control system in super critical one through boiler, *UTOPIA Conference*, Washington, Oct. 1999.

127. P. Matzinger, Tolerance, danger and the extended family, *Ann. Rev. Immunol.*, vol. 12, pp. 991–1045, 1994.

128. K.-C. Kwak, and D.-H. Kim, Adaptive neuro-fuzzy networks with the aid of fuzzy granulation, *IEICE Trans. Information Syst.*, vol. E88-D, no. 9, pp. 2189–2196, January 2005.

129. P. Matzinger, An innate sense of danger, *Semin. Immunol.*, vol. 10, pp. 399–415, 1998.
130. D. Moskophidis, F. Lechner, H. Pircher, and R. M. Zinkernagel, Virus persistence in acutely infected immuno competent mice by exhaustion of antiviral cytotoxic effector T-cells, *Nature*, vol. 362, pp. 758–761, 1993.
131. M. Oprea and S. Forrest, How the immune system generates diversity: Pathogen space coverage with random and evolved antibody libraries, *GECCO 99*, Real-World Applications Track, 1999.
132. D. H. Kim, et al., Context-based Fuzzy clustering for rule extraction of Takagi-Sugeno fuzzy model, *J. Konwledge-Based Systems* (SCI), April, 2004. pp.
133. D. G. Osmond, The turn over of B-cell populations, *Immunol. Today*, vol. 14, no. 1, pp. 34–37, 1993.
134. R. Eberchart and J. Kennedy, A new optimizer using particle swarm theory, *Proceedings of the International Symposium on Micro Machine and Human Science* (Nagoya, Japan), pp. 39–43, Oct. 1995.
135. W. E. Paul, *Fundamental Immunology*, 2nd ed., Raven Press Ltd., 1989.
136. W. K. Potts, C. J. Manning, and E. K. Wakeland, Mating patterns in semi natural populations of mice influenced by MHC genotype, *Nature*, vol. 352, pp. 619–621, 1991.
137. D H. Kim, The comparison of characteristics of 2-DOF PID controllers and intelligent tuning for a gas turbine generating plant, *Lecture Notes in Computer Science Proceeding of Springer* (SCI) Sept. 22–24, 2004.
138. D. H. Kim, Robust PID controller tuning using multiobjective optimization based on clonal selection of immune algorithm, *Lecture Notes in Computer Science Proceeding of Springer* (SCI) Sept. 22–24, 2004.
139. P. J. Fleming and R. C. Purshouse, Evolutionary algorithms in control system engineering: A survey, *Control. Eng. Prac.*, vol. 10, pp. 1223–1241, 2002.
140. M. Dotoli, G. Maione, D. Naso, and E. B. Turchiano, Genetic identification of dynamical systems with static nonlinearities, *Proceedings of the IEEE SMCia/01*, Mountain Workshop Soft Computing Industrial Applications, Blacksburg, VA, pp. 65–70, June 25–27, 2001.
141. G. J. Gray, D. J. Murray-Smith, Y. Li, K. C. Sharman, and T. Weinbrenner, Nonlinear model structure identification using genetic programming, *Contr. Eng. Pract.*, no. 6, pp. 1341–1352, 1998.

142. Kristinnson and G. A. Dumont, System identification and control using genetic algorithms, *IEEE Trans. System, Man, Cybern.*, vol. 22, pp. 1033–1046, Sept.-Oct. 1992.

143. B. Maione, D. Naso, and B. Turchiano, GARA: A genetic algorithm with resolution adaptation for solving system identification problems, *Proceedings of the European Control Conference* (Porto, Portugal), pp. 3570–3575, Sept. 4–7, 2001.

144. C. M. Fonseca and P. J. Fleming, Multiobjective optimization and multiple constraint handling with evolutionary algorithms-Part I: A unified formulation; "Part II: Application example," *IEEE Trans. Syst. Man Cybern. A: Syst. Humans*, vol. 28, no. 1, pp. 26–47, Jan. 1998.

145. K.-C. Kwak and D.-H. Kim, TSK-based Linguistic Fuzzy Model with uncertain model output, *IEICE Trans. Inf. Syst.*, vol. E86-D, No. 1 March 20, DOI: 10.1093/ietisy/e89-d.12.2919, 2005.

146. T. Glad and L. Ljung, *Control Theory, Multivariable and Nonlinear Methods*, London, U.K, Taylor & Francis, 2000.

147. D. H. Kim, J. H. Cho, *Robust Tuning of PID Controller with Disturbance Rejection Using Bacterial Foraging Based Optimization*, International Fuzzy System Association, Tsinghua University, Sichuan University, Aug. 28–31, 2005.

148. D. H. Kim, J. H. Cho, Optimal learning of fuzzy neural network using particle swarm optimization algorithm, *ICCASE2005*, Bucheon, Korea June 2–5, 2005.

149. Takahashi and T. Noguchi, A new quick response and high efficiency control strategy of an induction motor, *Conference Record IEEE-IAS Annual Meeting*, 1985, pp. 495–502.

150. A. M., Trzynadlowski, *Control of Induction Motors*, Academic Press, 2001.

151. M. Depenbrock, Direct self control for high dynamics performance of inverter feed a.c. machines, *ETZ Arch.*, vol. 7, no. 7, pp. 211–218, 1985.

152. Boldea and S. A. Nasar, Torque vector control (TVC)—A class of fast and robust torque, speed and position digital controllers for electric drives, *Conference Record EMPS'88*, vol. 15, 1988, pp. 135–148.

153. D. Casadei, F. Profumo, G. Serra, and A. Tani, FOC and DTC: Two viable schemes for induction motors torque control, *IEEE Trans. Power Electron.*, vol. 17, pp. 779–787, Sept. 2002.

154. P. Pohjalainen, P. Tiitinen, and J. Lalu, The next-generation motor control method—Direct torque control, DTC, *Conference Record EPE'94*, 1994, pp. 115–120.

155. J.-K. Kang and S.-K. Sul, New direct torque control of induction motor for minimum torque ripple and constant switching frequency, *IEEE Trans. Ind. Appl.*, vol. 35, pp. 1076–1082, Sept./Oct. 1999.

156. D. Casadei, G. Serra, and A. Tani, Implementation of a direct torque control algorithm for induction motors based on discrete space vector modulation, *IEEE Trans. Power Electron.*, vol. 15, pp. 769–777, July 2000.

157. T. G. Habetler, F. Profumo, M. Pastorelli, and L. M. Tolbert, Direct torque control of induction machines using space vector modulation, *IEEE Trans. Ind. Appl.*, vol. 28, pp. 1045–1053, Sept./Oct. 1992.

158. B. H. Kenny and R. D. Lorenz, Stator and rotor flux based deadbeat direct torque control of induction machines, *Conference Record IEEE-IAS Annual Meeting*, vol. 1, 2001, pp. 133–139.

159. C. Lascu, I. Boldea, and F. Blaabjerg, A modified direct torque control for induction motor sensorless drive, *IEEE Trans. Ind. Appl.*, vol. 36, pp. 122–130, Jan./Feb. 2000.

160. Mir, M. E. Elbuluk, and D. S. Zinger, Fuzzy implementation of direct selfcontrol of induction motors, *IEEE Trans. Ind. Appl.*, vol. 30, pp. 729–735, May/June 1994.

161. P. Z. Grabowski, M. P. Kazmierkowski, B. K. Bose, and F. Blaabjerg, A simple direct-torque neuro-fuzzy control of PWM-inverter-fed induction motor drive, *IEEE Trans. Ind. Electron.*, vol. 47, pp. 863–870, Aug. 2000.

162. V. Utkin, J. Guldner, and J. Shi, *Sliding Mode Control in Electromechanical Systems*, New York: Taylor & Francis, 1999.

163. Z. Yan, C. Jin, and V. I. Utkin, Sensorless sliding-mode control of induction motors, *IEEE Trans. Ind. Electron.*, vol. 47, pp. 1286–1297, Dec. 2000.

164. C. Lascu and A. M. Trzynadlowski, Combining the principles of sliding mode, direct torque control and space vector modulation in a high performance sensorless AC drive, *Conference Record IEEE-IAS Annual Meeting*, vol. 3, 2002, pp. 2073–2078.

165. H.-J. Shieh and K.-K. Shyu, Nonlinear sliding-mode torque control with adaptive back stepping approach for induction motor drive, *IEEE Trans. Ind. Electron.*, vol. 46, pp. 380–389, Apr. 1999.

166. M. Rodic and K. Jezernik, Speed-sensorless sliding-mode torque control of an induction motor, *IEEE Trans. Ind. Electron.*, vol. 49, pp. 87–95, Feb. 2002; A. Benchaib, A. Rachid, E. Audrezet, and M. Tadjine, Real-time sliding-mode observer and control of an induction motor, *IEEE Trans. Ind. Electron.*, vol. 46, pp. 128–137, Feb. 1999.

167. J.-O. Krah and J. Holtz, High performance current regulation and efficient PWM implementation for low-inductance servo motors, *IEEE Trans. Ind. Appl.*, vol. 35, pp. 1039–1049, Sept./Oct. 1999.

168. Poirier, E., Ghribi, M., and Kaddouri, A., Loss minimization control of induction motor drives based on genetic algorithms, *IEMDC 2001, IEEE Int.*, pp. 475–478, 2001.

169. CHEE-MUN ONE, *Dynamic Simulation of Electric Machinery*, Prentice Hall PTR, 1998.

170. Bolden, S. A. Nasar, *Electric Drives*, CRC Press, 1999.

171. D. H. Kim, J. H. Cho, Robust tuning of PID controller using bacterial-foraging-based optimization, *JACIII*, vol. 9, no. 6, pp. 669–676, 2005.

172. C. H. Lee and C. C. Teng, A novel robust PID controllers design by fuzzy neural network, *Asian J. Control.*, vol. 4, no. 4, pp. 433–438, 2002.

173. X. Xu, Y. M. Pok, C. Liu, and C. C. Hang, Tuning and analysis of a fuzzy PI controller based on gain and phase margins, *IEEE Trans. Syst. Man Cybern.-Part A: Syst. Humans*, vol. 28, no. 5, pp. 685–691, 1998.

174. T. W. Kraus, and T. J. Mayron, Self-tuning PID controllers based on a pattern recognition approach, *Control. Eng. Pract.*, 106–111, 1984.

175. C. C. Hang, W. K. Ho, and L. S. Cao, A comparison of two design methods for PID controllers, *ISA Trans.*, vol. 33, 147–151, 1994.

176. M. Zhuang and D. P. Atherton, Automatic tuning of optimum PID controllers, *IEE Proc. Part D*, vol. 140, no. 3, 216–224, 1993.

177. J. E. Burke, Scattering of surface waves on an infinitely deep fluid, *J. Math. Phys.*, vol. 6, pp. 805–819.

178. D. H. Kim, P. Baranyi, Novel emotion dynamic express for robot, sami2011, Slovakia.

179. D. H. Kim, Emotion Control of Induction Motor Using Disturbance Acceptance and Bacteria Foraging, Proceedings of the 2013 International Conference on Artificial Intellige NCE, July 22–25, 2013 Las Vegas Nevada, USA.

180. D. H. Kim, S. B. Nair, Korean Brain Pool Report, 2008.

181. D. H. Kim, Design and tuning approach of 3-DOF emotion intelligent PID (3-DOF-PID) controller, *IEEE-EMS2013 Conference*, Nov. 11–12, 2012. Malter, 2012.

182. D. H. Kim, S. B. Nair, Novel emotion engine for robot and its parameter tuning by bacterial foraging, *SACI2009*, Hungary.

183. D. H. Kim, Fuzzy rule based voice emotion control for user demand speech generation of emotion robot, *ICCAT2013*, Tunisia.

184. R. W. Picard, *Future Interfaces: Social and Emotional*, MIT Media Lab, 2002.

185. R. W. Picard, Toward machine emotional intelligence: Analysis of affective physiological state, *IEEE Tans. Pattern Anal. Mach.*, vol. 23, no. 10, October 2001.

186. L. Lungu, Artificial emotion simulation model, http://aimas.cs.pub.ro/people/valentin.lung.

187. Hossein Rouhani, Emotional learning based intelligent speed and position control applied to neurofuzzy model of switched reluctance motor, *Control. Cybern.*, vol. 36, no. 1, 2007.

188. C. Becker, *Simulating the Emotion Dynamics of a Multimodal Conversational Agent*, Springer-Verlag Berlin Heidelberg, pp. 154–165, 2004.

189. J. Sidorova, Speech Emotion Recognition, DEA report 2007. KR.

190. K. R. Scherer, Expression of emotion in voice and music, *J. Voice*, vol. 9, no. 3, 235–248, 1995.

191. J. Yu, et al. Brief Paper, Comparison of nonlinear control design techniques on a model of the Caltech ducted fan, *Automatica*, vol. 37, pp. 1971–1978, 2001.

192. G. A. Sohl, J. E. Bobrow, Experiments and simulations on the nonlinear control of a hydraulic servo system, *Trans. Control Systems TECH.*, vol. 7, no. 2, pp. 238–247, 1999.

193. B. Yao, Adaptive robust control of nonlinear systems with application to control of mechanical systems, Thesis, UCB, USA, 1996.

194. A. U. Levin, et al., Control of nonlinear dynamic systems using neural network: controllability and observability, *IEEE Trans. Neural Netw.*, pp. 192–206, 1993.

195. C. C. Lee, Fuzzy logic in control system: Fuzzy logic controller, part I and II, *IEEE Trans. Syst. Man. Cybern.*, vol. 20, no. 2, pp. 404–435, 1990.

196. S. Park, I. W. Sandberg, Approximation and radial-basis-function networks, *Neural Computer*, pp. 105–110.

197. H. Wang, M. Brown, C. J. Harris, Neural network modeling of unknown nonlinear systems subject to immeasurable disturbances, *IEE Proc., Control Theory Appl.*, vol. 141, no. 4, pp. 216–222, 1994.

198. S. Horikawa, T. Furuhashl, Y. Uchikawa, On fuzzy modeling using fuzzy neural networks with back propagation algorithm, *IEEE Trans. Neural Netw.*, pp. 801–806, 1992.

199. W. A. Farag, V. H. Quintana, G. Lambert-Torred, A genetic-based neuro-fuzzy approach for modeling and control of dynamical systems, *IEEE Trans. Neural Netw.*, vol. 9, no. 5, Sept. 1998.

200. K. Mori, M. Tsukiyama, Immune algorithm with searching diversity and its application to resource allocation problem, *Trans. JIEE*, vol. 113-C, no. 10, 1993.

201. D. H. Kim, Intelligent tuning of a PID controller using an immune algorithm, *Trans. KIEE*, vol. 51-D, no. 1, 2002.

202. D. H. Kim, PID controller tuning of a boiler control system using immune algorithm typed neural network, ICCS2004, Poland, June 2004.

203. B. J. Park, Fuzzy polynomial neural networks: Hybrid architectures of fuzzy modeling, *IEEE Trans. Fuzzy Systems*, vol. 10, no. 5, pp. 607–621, Oct. 2002.

204. B. S. Ko, Application of intelligent controller to industry, *KFIS Spring Conf.*, vol. 11, pp. 22–25, 1997.

205. J.-S. R. Jang, C.-T. Sun, E. Mizutani, *Neuro-Fuzzy and Soft Computing*, Prentice-Hall, 1997.

206. G. A. Carpenter, S. Grossberg, *Competitive Learning: From Interactive Activation to Adaptive Resonance*, chapter 5, pp. 213–250. MIT Press, Cambridge, 1987.

207. G. A. Carpenter, S. Grossberg, A massively parallel architecture for a self-organizing neural pattern recognition machine, *Comput. Vision Graph. Image Process.*, vol. 37, pp. 54–115, 1987.

208. T. P. Caudell, D. S. Newman, An adaptive resonance architecture to define normality and detect novelties in time series and databases, *IEEE World Congress on Neural Network*, Portland, Oregon, pp. IV166–176, July 3–7, 1993.

209. D. Chowdhury, D. Stauffer, Statistical physics of immune networks, *Physica A*, vol. 186, pp. 61–81, 1992.

210. D. Dasgupta, Immunity-based systems, a survey. Presented at the *First Workshop on Immunity-Based Systems (IMBS'96)*, Nara, Japan, Dec. 10, 1996.

211. D. Dasgupta, Using immunological principles in anomaly detection, *Proceedings of the Artificial Neural Networks in Engineering (ANNIE'96)*, St. Louis, USA, Nov. 10–13, 1996.

212. D. Dasgupta, S. Forrest, Novelty detection in time series data using ideas from immunology, *ISCA 5th International Conference on Intelligent Systems*, Reno, Nevada, June 19–21, 1996.

213. P. D'haeseleer, S. Forrest, P. Helman, An immunological approach to change detection: Algorithms, analysis, and implications, *Proceedings of IEEE Symposium in Research in Security and Privacy*, Oakland, CA, May 1996.

214. J. D. Farmer, A Rosetta Stone for connectionism, *Physica D*, vol. 42, pp. 153–187, 1990.

215. S. Forrest, A. S. Perelson, L. Allen, R. Cherukuri, Self-nonself discrimination in a computer, *Proceedings of IEEE Symposium in Research in Security and Privacy*, pp. 202–212, Oakland, CA, May 16–18, 1994.

216. S. A. Frank. *The Design of Natural and Artificial Adaptive Systems*. Academic Press, New York, M. R. Rose and G. V. Lauder edition, 1996.

217. L. Fu. *Neural Networks in Computer Intelligence*. McGraw-Hill, Inc., 1st ed., 1994.

218. G. W. Hoffmann, A neural network model based on the analogy with the immune system, *J. Theor. Biol.*, vol. 122, pp. 33–67, 1986.

219. J. E. Hunt, D. E. Cooke, Learning using an artificial immune system, *J. Netw. Comput. Appl.*, vol. 19, pp. 189–212, 1996.

220. A. S. Perelson, Immune network theory, Immunol. Rev., vol. 10, pp. 5–36, 1989.

221. A. S. Perelson, G. Weisbuch, Immunology for physicists, Preprint for *Review of Modern Physics*, June 1995 .

222. G. W. Rowe, *The Theoretical Models in Biology*, Oxford University Press, 1st ed., 1994.

223. V. Vemuri, Artificial neural networks in control applications, *Adv. Comput.*, vol. 36, pp. 203–254. Academic Press, Inc., M. Yovits edition, 1993.

224. F. T. Vertosick, R. H. Kelly, Immune network theory: A role for parallel distributed processing?, *Immunology*, vol. 66, pp. 1–7, 1989.

225. F. T. Vertosick, R. H. Kelly, The immune system as a neural network: A multi-epitope approach, *J. Theor. Biol.*, vol. 150, pp. 225–237, 1991.

226. M. Kawski, Intro to Lyapunov theory of stability, *APM 581*, Nov. 15, 2009.

227. X. Chen, et al., Design of a nonlinear disturbance observer, *IEEE Trans. Ind. Electron.*, vol. 47, no. 2, pp. 429–437, April 2000.

228. R. A. Krohling, J. P.Rey, Design of optimal disturbance rejection PID controllers using genetic algorithms, *IEEE Trans. Evolut. Comput.*, vol. 5, no. 1, Feb. 2001.

229. C.-K. Weng, A. Ray, Modeling of power plant dynamics and uncertainties for robust control synthesis, *Appl. Math. Modeling*, vol. 20, pp. 501–512, May 1996.

230. A. Ishiguro, T. Kondo, Y. Watanabe, Y. Uchikawa, Dynamic behavior arbitration of autonomous mobile robots using immune networks, *Proc. ICEC' 95*, vo1. 2, pp. 722–727, 1995.

231. D. H. Kim, J. H. Cho, Design of robust PID controller with disturbance rejection for motor using immune algorithm, *Hybrid Int. Sys. Conference, HIS 2004*, Kitakyushu, Japan, 2004.

232. J. D. Farmer, N. H. Packard, A. S. Perelson, The immune system, adaptation, and machine learning, *Physica D*, vol. 22, pp. 187–204, 1986.

233. R. Brooks, A robust layered control system for a mobile robot, *IEEE J. Robot. Autom.*, vol. 2, pp. 14–23, 1988.

Index